T0291872

CAMBRIDGE LIBRARY COLLECTION

Books of enduring scholarly value

Travel and Exploration

The history of travel writing dates back to the Bible, Caesar, the Vikings and the Crusaders, and its many themes include war, trade, science and recreation. Explorers from Columbus to Cook charted lands not previously visited by Western travellers, and were followed by merchants, missionaries, and colonists, who wrote accounts of their experiences. The development of steam power in the nineteenth century provided opportunities for increasing numbers of 'ordinary' people to travel further, more economically, and more safely, and resulted in great enthusiasm for travel writing among the reading public. Works included in this series range from first-hand descriptions of previously unrecorded places, to literary accounts of the strange habits of foreigners, to examples of the burgeoning numbers of guidebooks produced to satisfy the needs of a new kind of traveller - the tourist.

A Summer in Greenland

The author of the standard early twentieth-century textbook on fossil plants, A. C. Seward (1863–1941) was Professor of Botany at Cambridge, Master of Downing College and Vice-Chancellor of Cambridge University. This account of his first research trip to Greenland is an evocative portrait of the country, its immense and sublime landscape, its people, and life on the Danish scientific station. This little book, written in an engaging conversational tone, conveys Seward's enthusiasm for Greenland. It includes an explanation for non-specialists of some of Seward's findings relating to fossil plants found there, which provide evidence that the country had a much milder climate in previous geological periods. Seward's own photographs are a fascinating record of the traditional life of the Inuit population as it then survived, as well as the rugged scenery of icebergs and glaciers.

Cambridge University Press has long been a pioneer in the reissuing of out-of-print titles from its own backlist, producing digital reprints of books that are still sought after by scholars and students but could not be reprinted economically using traditional technology. The Cambridge Library Collection extends this activity to a wider range of books which are still of importance to researchers and professionals, either for the source material they contain, or as landmarks in the history of their academic discipline.

Drawing from the world-renowned collections in the Cambridge University Library, and guided by the advice of experts in each subject area, Cambridge University Press is using state-of-the-art scanning machines in its own Printing House to capture the content of each book selected for inclusion. The files are processed to give a consistently clear, crisp image, and the books finished to the high quality standard for which the Press is recognised around the world. The latest print-on-demand technology ensures that the books will remain available indefinitely, and that orders for single or multiple copies can quickly be supplied.

The Cambridge Library Collection will bring back to life books of enduring scholarly value (including out-of-copyright works originally issued by other publishers) across a wide range of disciplines in the humanities and social sciences and in science and technology.

A
Summer in
Greenland

A.C. SEWARD

CAMBRIDGE
UNIVERSITY PRESS

CAMBRIDGE UNIVERSITY PRESS

Cambridge, New York, Melbourne, Madrid, Cape Town, Singapore,
São Paolo, Delhi, Dubai, Tokyo, Mexico City

Published in the United States of America by Cambridge University Press, New York

www.cambridge.org
Information on this title: www.cambridge.org/9781108012836

This edition first published 1922
This digitally printed version 2010

ISBN 978-1-108-01283-6 Paperback

A SUMMER IN GREENLAND

CAMBRIDGE UNIVERSITY PRESS

C. F. CLAY, Manager

LONDON : FETTER LANE, E.C. 4

NEW YORK : THE MACMILLAN CO.
BOMBAY }
CALCUTTA } MACMILLAN AND CO., LTD.
MADRAS }
TORONTO : THE MACMILLAN CO. OF
CANADA, LTD.
TOKYO : MARUZEN-KABUSHIKI-KAISHA

PLATE I

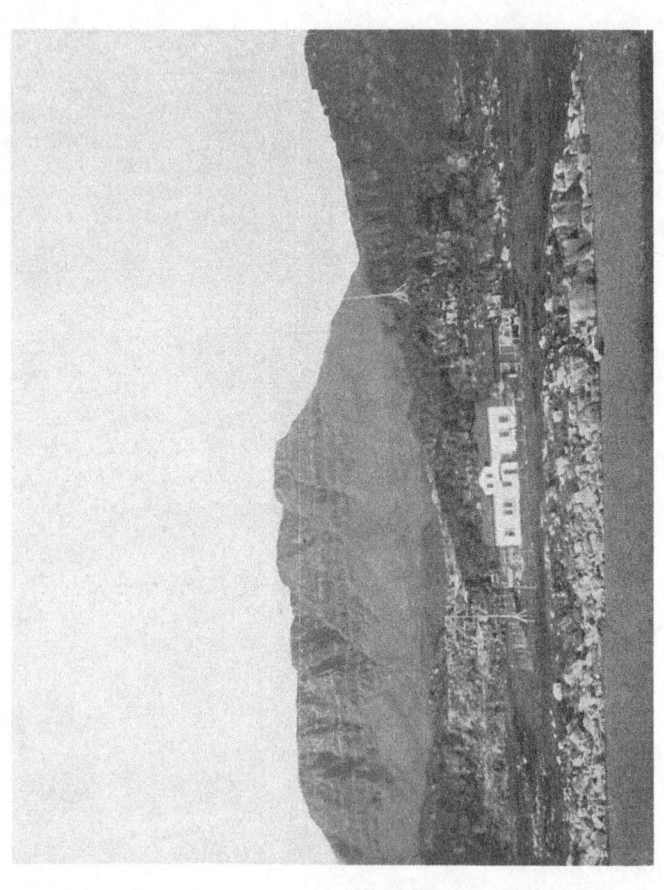

THE DANISH ARCTIC STATION, DISKO ISLAND, OVERLOOKING DISKO BAY

A Summer in Greenland

BY

A. C. SEWARD

MASTER OF DOWNING COLLEGE
AND PROFESSOR OF BOTANY IN THE
UNIVERSITY OF CAMBRIDGE

CAMBRIDGE
AT THE UNIVERSITY PRESS
1922

TO
MY WIFE

PREFACE

A GOOD many years ago British Arctic Explorers, on their way to the Polar Sea, and other travellers first brought to Europe collections of fossil plants from Disko Island and from various localities on the adjacent coasts of Greenland. In more recent years large collections, made by Danish and Swedish Geologists, have been acquired by the museums of Copenhagen and Stockholm.

To students of the vegetation of the past fossil plants from Greenland rocks are of exceptional interest, mainly because of the evidence they afford of climatic conditions very different from those within the Arctic Circle at the present day. An examination of the Copenhagen and Stockholm collections inspired me with a desire to visit Greenland in order to obtain as representative a set of fossils as possible for the British Museum and for Cambridge. Having learnt from my friend Professor Ostenfeld (of Copenhagen) that English visitors would be welcome at the Danish Arctic Station on Disko Island, I applied through the British Foreign Office for the necessary permission to visit the country, and, with the help of a grant from the Royal Society, supplemented by a grant from the Cambridge University Worts Travelling Fund, I was at length able to gratify my wish. Had it not been for the assistance generously given to me by Professor Ostenfeld my desire would in all probability have remained unfulfilled.

Mr R. E. Holttum of St John's College, who has recently been appointed Assistant Director of the Singapore Botanic Garden, accompanied me as Research Assistant and fully justified the high opinion I had previously formed of his keenness and ability as a botanist. I am grateful to him for many services in addition to those of a strictly scientific nature. To Mr Daugaard-Jensen, the Director for Greenland, and to many Danish officials resident in Greenland I am indebted for much willingly-rendered assistance. In a land where there are no hotels or guest-houses a foreign visitor is necessarily dependent on private hospitality, and this was generously extended to us. My indebtedness to Mr Morten Porsild, the Director of the Danish Arctic Station, cannot be adequately expressed.

In the following pages, the publication of which I will not attempt to justify, my aim has been to avoid technical details as far as possible and to confine myself to a general, and necessarily very incomplete, treatment of the botany and geology of the country; but as the object of the journey was scientific preference is given to natural history subjects. I have not attempted to deal other than very superficially with the history of Greenland, with the present system of government, or with the life of the people. My own observations and impressions have been supplemented by facts obtained from some of the excellent contributions to the *Meddelelser om Grønland* (Communications on Greenland), more than fifty volumes of which have already appeared: this publication, devoted exclusively to subjects connected with Greenland, reflects the greatest credit upon the Danish Government. Free use has also been made of information given to me in conversation by Mr Porsild.

Passages from the botanical and geological portions of this sketch and a few of the illustrations have been published in *Discovery*: for permission to reproduce them here my thanks are due to the Publisher and Editor of that Journal.

The celebration in July of last year of the Bicentenary of the landing of Hans Egede in Greenland was not only made the occasion of a royal visit, but the event was marked by the publication, at the end of 1921, of a comprehensive work by several authors, entitled *Grønland*, written in Danish, and prepared under the auspices of the Commission for promoting scientific investigation in the country.

Though it would be out of place here to enumerate the species of recent plants collected, I take this opportunity of thanking the following botanists for examining and determining our material: Miss A. Lorrain Smith (lichens), Mr H. N. Dixon (mosses), Mr W. E. Nicholson (liverworts), and Dr W. J. Hodgetts (freshwater algae). Nearly 200 species of flowering plants were obtained, chiefly by Mr Holttum: a representative collection of these has been sent to the Herbarium, the Royal Gardens, Kew; others have been given to the British Museum and to the Botany School, Cambridge. Rock specimens were divided between the British Museum and Cambridge. The majority of the fossil plants, of which nearly 1000 specimens were obtained, will be sent to the British Museum after their description has been completed; the rest will be kept at Cambridge.

I am grateful to my friends Mr Peace, the University printer, and Mr S. C. Roberts, Secretary to the Syndics of the Press, for many helpful suggestions and for the interest they have taken in the publication of this book.

I am indebted to my wife for the drawings reproduced in Figs. 12 and 46, which were made from my rough sketches, and to my daughter Phyllis for Map *B* and the Chart reproduced on page 24. To Mr Holttum I am indebted for the photographs reproduced in Figs. 9, 18, 22, 25–31, 44, and to Mr Erling Porsild for Figs. 23 and 43. Fig. 1 is from a photograph taken by the Greenland photographer, John Moller, of Godthaab. With these exceptions the figures are from negatives taken by myself with a Kodak camera.

A. C. SEWARD

August 5, 1922

CONTENTS

CHAPTER I

CHAPTER II

CHAPTER III

CHAPTER IV

CHAPTER V

CHAPTER VI

PLATES

MAPS

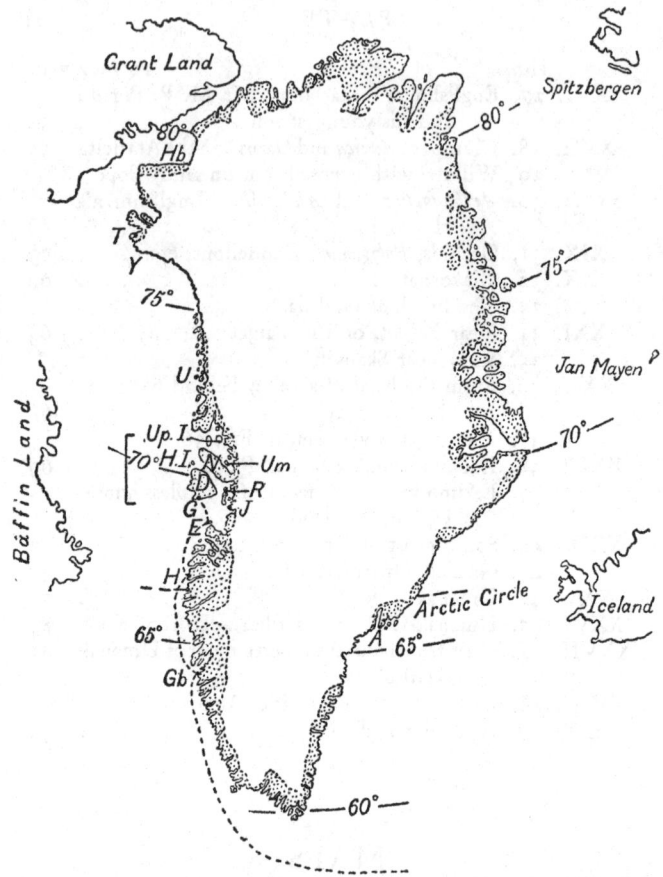

MAP *A*. Greenland. (Scale approximately, 1 inch = 425 miles.) The dotted coast-regions are for the most part free from ice in the summer; the rest is under the inland ice.

A. Angmagssalik (east coast S. of the Arctic Circle). D. Disko Island (west coast, lat. 70° N.). E. Egedesminde (west coast, lat. 68° N.). G. Godhavn, Disko Island. Gb. Godthaab (west coast, lat. 64° N.). H. Holsteinsborg (west coast, near Arctic Circle). Hb. Humboldt glacier (lat. 80° N.). Hare Island (N.W. of Disko Island). J. Jakobshavn (lat. 69° N.). N. Nûgssuaq Peninsula (lat. 70° N.). R. Ritenbenk (lat. 70° N.). T. Thule (lat. 77° N.). U. Upernivik (lat. 73° N.). Up. I. Upernivik Island (lat. 71° N.). Um. Umanak (off the east coast of the Nûgssuaq Peninsula). Y. Cape York (south of Thule).

The broken line shows the route as far as Disko Island.

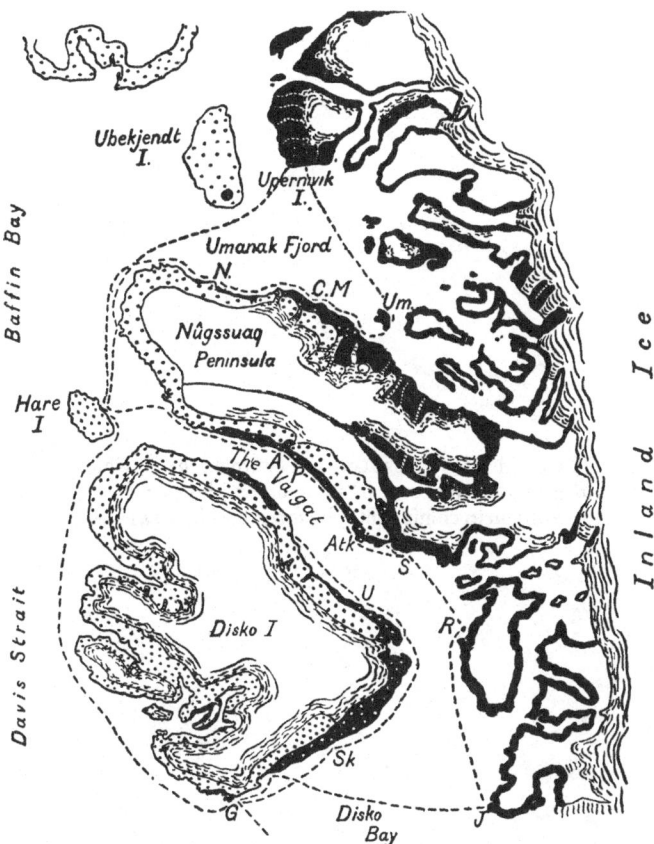

Map B. (Scale approximately, 1 inch = 36 miles.) Map showing on a larger scale the area within the square bracket on the west side of Map A. Gneiss and other igneous rocks belonging to the oldest geological period, the Archaean, are indicated so far as their areas have been surveyed, by solid black. Basalts and volcanic ash are indicated by white with black dots; sedimentary rocks by black with white dots. The edge of the inland ice is shown by wavy black lines.

Most of the places visited and mentioned in the text are marked on the map. Motor boat routes thus ------. A. Ata. Atk. Atanikerdluk. C.M. Government Coal-mine. G. Godhavn. J. Jakobshavn. N. Niakornat. P. Patoot. R. Ritenbenk. S. Sarkak. Sk. Skansen. U. Ujaragsugssuk. Um. Umanak.

"The child looks straight upon Nature as she is, while a man sees her reflected in a mirror, and his own figure can hardly help coming into the foreground." LATHAM.

CHAPTER I

Though Greenland be a Country of a vast extent, yet it affords but a narrow field for any observation or remarks of consequence; there being no strong or well-built towns to meet with; no well-ordered Polity or Civil Government; no fine Arts and Sciences, or the like; but only a number of mean, wretched, and ignorant Gentiles, who live and improve the land according to their low capacity.   HANS EGEDE, 1741.

The colonisation of Greenland by Eric the Red; its re-colonisation by Hans Egede in 1721. From Copenhagen to Disko Island in 1921. The Danish Arctic Station. The last resting-place of the 'Fox.'

IN the latter part of the tenth century there lived in Iceland a 'courageous, indomitable, and quarrelsome man' called Eric the Red, who had come there as a child from Norway with his outlawed father. Having in his turn been declared an outlaw, Eric about the year 983 equipped a ship and set sail for a land which had recently been discovered by one Gunbjörn to the west of Iceland. After sighting the south-east coast of the newly-discovered country he rounded the southern headland, christened Cape Farewell some centuries later by John Davis, and landed on the south-west coast. Eric soon returned to Iceland and, believing that a 'comely' name would induce others to throw in their lot with him, he called the country Greenland. The result was, thirty-five ships left Iceland and fourteen of them reached their destination.

The wooden shells of old Scandinavian ships discovered in modern times enable us more thoroughly to appreciate the courage of these pioneers of colonisation. An unusually good representation of a Viking's ship, which shows the rigging more clearly than on the ships depicted on the Bayeux Tapestry, may be seen on a large incised slab of limestone in the Museum at the famous Hanseatic town of Visby in the Island of Gothland.

Many farms were established by the colonists and stocked with cattle brought from Iceland. Cattle, especially sheep, are still kept in the more southern parts of Greenland, but in the northern districts, where sledges are used in the winter and dogs are essential, it is practically impossible, apart from the difficulty of providing food, to keep domestic animals. A few Danish residents keep goats for the luxury of having fresh milk, but these are a cause of anxiety because of the difficulty of protecting them against the attacks of hungry dogs. South of Holsteinsborg (the southernmost place where sledge-dogs are kept—see Map *A*, H) goats and chickens are common. The dogs which are kept in South Greenland for the sake of their skins are said to be very tame as compared with the sledge-dogs of the north.

The discovery of a Runic stone a little to the north of lat. 72° N.[1] affords evidence of the long

---

[1] A fuller account of this and other facts connected with the history of Greenland is given by Sir Clements Markham in *The Lands of Silence*, Cambridge, 1921. For further details the reader is referred to papers in the *Meddelelser om Grønland* and to the recently-published book (in Danish) entitled *Grønland*, Copenhagen, 1921.

distances travelled by the Norse colonists presumably in search of seals and other animals. Christianity was introduced soon after the arrival of the first colonists; the first of a long line of bishops was appointed, and churches were built. In 1261 the Republic of Greenland became part of the powerful Norwegian kingdom, which also included the Orkney and Shetland Islands, the Faroes, Iceland, and the Isle of Man. For a time the new colony flourished, but about the middle of the fourteenth century, owing to various causes, communication with the home country practically ceased and a period of decline ensued. It was probably in the thirteenth century that the Norsemen first came into contact with the Eskimoes, or Skraellings as they were then called. Accounts have been handed down of fights between the natives and the Icelanders and excavations have brought to light Norse skulls transfixed by stone arrow-heads of a type that is still found among the débris of old Eskimo settlements. Though precise information is lacking, it is probable that the Norsemen ceased to exist in Greenland more than three hundred years ago. When Martin Frobisher and John Davis landed in Greenland in 1578 and 1585 respectively, no Norsemen were seen.

In the early part of the eighteenth century Hans Egede, a pastor in Norway and the son of a Dane, was able to gratify his desire to visit Greenland in search of descendants of the original settlers. He sailed from Bergen on May 3, 1721, in the 'Haabe' (the 'Hope'). Hans Egede felt that it would be 'his greatest happiness and joy' to be allowed to preach Christianity to his neglected and demoralised countrymen. On July 3 the 'Hope' anchored off an island near the flourishing Settlement of Godthaab (Map *A*, Gb; lat. 64° N.[1]). His hope of finding survivors was not realised; he found only ruins of their houses and churches. Many relics of the Norse civilisation have been discovered in different parts of South Greenland and archaeological research is still in progress. I learnt from the Inspector for South Greenland that some important new discoveries had been made last summer (1921) by Danish archaeologists, notably the remains of Norse skeletons with complete and well-preserved clothing. The following extract from an American source, quoted by Mr Erling Porsild in an article recently published in a Danish newspaper, gives a more picturesque account of the recent discoveries in South Greenland: '*Entombed in an Iceberg for a Thousand Years*. The startling discovery which came when the Grand Floe drifted ashore and revealed a picturesque Viking, perfectly preserved, even to his winged helmet, armor and red hair.' The Viking was presumably Eric the Red!

Though disappointed in his main objective, Hans Egede determined to devote himself to the education and conversion of the natives while he also engaged in trade. This he did for fifteen years when he returned to Europe, leaving his son Paul to continue the good work.

[1] The degrees of latitude are stated only with approximate accuracy to facilitate reference to the sketch-map.

PLATE II

John Voller (Godthaab) phot

FIG. I. GODTHAAB

PLATE III

FIG. 2. THE ANTLERS (HJORTETAKKEN), NEAR GODTHAAB

FIG. 3. THE S.S. 'HANS EGEDE' AT EGEDESMINDE

In 1814, on the separation of Denmark from Norway, the greater part of Greenland became a Danish colony; and now the whole belongs to Denmark. For the first time in its history Greenland was visited in July of last year by the King and Queen of Denmark on the occasion of the celebration of the bicentenary of the landing of Hans Egede and the re-colonisation of the country. The visitors received an enthusiastic welcome from the delighted Greenlanders. On the rare occasions when a ship from Denmark visits a Settlement the natives crowd about the hills overlooking the sea and by their shouts announce the important event. The arrival of His Majesty's ship was naturally the cause of a very special demonstration and of unusual excitement. On the way to Greenland the royal party called at Iceland: this led to some confusion between the two countries in reports of the tour. An English illustrated paper published a photograph purporting to represent a scene in Greenland; the Queen is seated in 'an island cart' and the King is mounted on a 'sturdy Greenland pony.' Greenland possesses neither carts nor ponies.

In order to provide for the large number of people, several of whom were Delegates of Societies and Institutions invited to take part in the festivities, the Danish Government chartered a special vessel, the S.S. 'Bele,' a Swedish ship of about 1600 tons, to supplement the usual restricted service from Copenhagen. It was my good fortune, in company with my companion and Research

Assistant, Mr R. E. Holttum, of St John's College, Cambridge, to be one of the passengers. A few days after leaving Copenhagen, in the middle of an informal concert, 'God save the King' was heartily sung for the benefit of the two English travellers, a touching illustration of the uniform kindness we invariably received. One of my many pleasant memories is the friendliness of the Swedish captain, who, after the loss of the 'Bele,' was one of my companions, in very different circumstances, in the hold of the 'Hans Egede' on the return voyage to Denmark. Leaving Copenhagen on June 18 we reached Godthaab (Fig. 1) at midnight on June 28. On June 26, in rain, fog and a turbulent sea, we were somewhere off Cape Farewell; the conditions, though trying to the navigating officer, to whom floating ice was an additional cause of anxiety, appealed to the imagination as a fitting introduction to the mysterious land. The view near the Settlement of Godthaab is dominated by the two mountains, Sadelen and Hjortetakken (the Saddle and the Antlers), nearly 4000 ft. high, which were first ascended by members of a Swiss Expedition in 1909. Hjortetakken, its gneissic rocks weathered into outlines recalling the Langdale Pikes in the English Lake District, is shown in the photograph (Fig. 2). We sailed from Greenland on September 6 and arrived at Copenhagen on September 24.

Going ashore for the first time, especially when lack of knowledge of a country gives it an air of mystery, causes a thrill of excitement and in such circumstances it is easy to follow the sound but often neglected advice of Francis Galton—'Interest yourself chiefly in the progress of your journey, and do not look forward to its end with eagerness.'

Godthaab, which is often spoken of as the capital of Greenland, has about 400 inhabitants. Unlike the more northerly Settlements it has no dogs; it lies south of the region where sledges are used in the winter. Godthaab is the headquarters of the Senior Inspector, the Inspector for South Greenland. Fig. 1, reproduced from a print taken by the Greenland photographer, John Moller, shows the landing-stage on the extreme left, also one of the very few roads in the country. The large wooden house in the foreground is the residence of the chief Danish official; to the right in the distance is the Seminary, the largest house in Greenland, and to the left the church with a wooden spire. There are more modern buildings here than in most of the Settlements and very few typical stone-and-turf Eskimo houses. Not far from the church is a simple stone monument to the memory of Jørgen Brønlund, a Greenland member of the Denmark Expedition of 1906–1908 to the north-east coast, who perished on a sledge journey: the last entry in his diary ran as follows: 'I arrived here at the time of a declining moon and can go no further owing to frost bites on the feet and to the darkness. The bodies of the two others lie in the middle of the fjord in front of the glacier.' Close to the coast a short distance from the Settlement is the old Station of the

Moravian Mission, the members of which left the country in 1900. The Mission had been established in Greenland about two centuries. We reached Holsteinsborg (Map *A*, H; just within the Arctic Circle) on June 30. The houses of the Settlement, which has about 300 inhabitants but no resident doctor, are scattered among the rocks of the foot-hills overlooking an excellent harbour. A few of us went ashore to botanise, and, after wandering a short distance over the hills and across patches of snow lying on the edge of the lower swampy ground, we discovered how easy it is to lose oneself completely in a trackless country only a mile or two from a Settlement. At a height of about 300 feet I picked up an almost perfect shell of a sea urchin, one of several found in similar situations; these had probably been dropped by sea birds.

The departure from a Settlement is accompanied by protracted leave-taking; a long interval elapses between the first intimation of sailing and the actual weighing of the anchor. Social as well as business affairs are conducted with deliberation; there is no indecent haste. In Greenland, it is sometimes said, 'one hurries slowly.'

On July 2 we arrived at Egedesminde (lat. 68° N.) where we left the 'Bele' to continue her voyage further north to Upernivik (lat. 73° N.), which was never reached. On Sunday, July 3, a special service was held in the church in celebration of Hans Egede's Day. The pastor, at whose house we were hospitably entertained, told me that since ordination he had lived twenty-six years in Greenland. On July 4 we availed ourselves of an invitation from the Inspector for North Greenland (Mr Lindow) to accompany him and Mrs Lindow in his motor-boat to Godhavn, on the south coast of Disko Island, which we reached in six hours. In crossing the bay we were often enveloped in a dense fog; as it gradually rolled away, retreating like a high wall of uniform height, the sun lit up emerging icebergs; in front of us were the flat-topped basaltic mountains of Disko resting on rounded hummocks of the older gneiss—a remnant of the ancient continent which had existed for countless ages before the overlying strata were poured as molten lava over its weather-worn surface.

Godhavn has about 150 inhabitants; the Settlement, which was a favourite place of call for British whalers, and is described by many Arctic explorers, has no resident doctor. If medical aid is required on any part of Disko Island in the summer a message is sent by a Greenlander in his kayak to the doctor at Egedesminde; in the winter the journey is made on a dog-sledge. We took up our quarters at the Danish Arctic Station (Frontispiece and Fig. 9), which was to be our base. The Station is situated about a mile from the harbour.

The primary object of our expedition was to collect fossil and living plants and other natural history specimens. The localities visited for scientific purposes are included within the square bracket on the side of the larger map (Map *A*), from Upernivik Island to the south coast of Disko

Island, and are shown more clearly on Map *B*. Some of the places we visited are not marked on Map *B*, but our motor-boat routes are indicated by the dotted lines. With the exception of Upernivik Island (a short distance north of lat. 71° N.) most of the places at which we landed in the course of the motor-boat trips were visited by members of the American Peary Arctic Expedition of 1897 for the collection of geological specimens and plants and animals. So far as I know, that is the last occasion in recent years prior to 1921 on which an expedition with the collection of fossils as the main objective visited West Greenland. A general account of the American Expedition was published in the *Bulletin of the Geological Society of America*, vol. ix, 1898.

It is perhaps not generally known that Greenland is a 'closed' country; the trade in skins, seal oil, eiderdown, fish, and other products is a Government monopoly and no foreigner or even Danes are allowed to go there unless they have some definite purpose in view which is considered satisfactory by the Director for Greenland. A few steamers and sailing ships go direct from Copenhagen in the summer, and of these the S.S. 'Hans Egede,' on which we returned from Greenland in September, is the best known. The ship is seen in Egedesminde harbour in Fig. 3. On embarkation everyone must produce a medical certificate signed by a Danish doctor. The 'Hans Egede,' being built for navigation in seas where ice is abundant, is without a bilge-keel and has a well-deserved reputation for rolling. The ship, if clumsy, is strong; the captain, who is a man of attractive personality and thoroughly at home in Arctic seas, also inspired confidence. There are small though comfortable state rooms for about twenty passengers. On the return voyage in September, the third of the four trips made each summer, nearly sixty passengers had to be provided for and temporary accommodation was made in the forehold to which access was gained through a hatch, requiring some practice to negotiate with reasonable comfort, and a long, steep ladder. It was a new and interesting experience to travel as cargo, but despite a succession of rough seas our obliging and acrobatic steward was always able to carry the food along the wave-swept deck. Lying in a bunk in the bow of a pitching ship—the thud and swish of the waves overhead, the shivering and creaking of the timbers, cups swinging noisily on their hooks, heavy bodies occasionally careering across the floor of the damp hold—these and other impressions received by a sleepily receptive mind made one feel that the traveller in a transatlantic 'floating hotel' misses something which gives point to the contrast of a sea voyage to the smoothness and peace of one's normal way of life. 'On certain ships,' it is written in an advertisement recently published by a well-known Line of Steamers, 'one can hardly appreciate the fact that one is at sea.' The 'Hans Egede' in this respect has the advantage—the sea is obviously there.

Some weeks after we left the S.S. 'Bele' at Egedesminde—news travels slowly in a country with no regular postal service and no telegrams— we heard that the ship had struck the rocks of a small island in foggy weather a few miles south of her destination, the most northerly Danish Settlement of Upernivik (Map *A*, U, lat. 73° N.). Fortunately the 'Bele,' unlike the regular Danish steamers, possessed a wireless installation and was able to communicate with the King's ship rather more than 200 miles further south, which speedily went to her assistance and took back to Denmark some of the passengers and crew who had meanwhile made themselves as comfortable as they could in improvised tents on the inhospitable rocky coast. The wreck of the 'Bele' may, it is hoped, lead to the establishment in the near future of some wireless stations on the Greenland coast and on the Government ships. The statement that there is no regular postal service, though true in the ordinary sense, needs some modification. There is a recognised scale of payment for natives who convey messages or carry mails to Settlements within reasonable distance of the ports of call of steamers from Copenhagen. Upernivik is usually spoken of as the most northerly Danish Settlement in Greenland, but there is a trading station between lat. 76° N. and 77° N., established in 1910 under the name Thule, by the well-known Danish explorer, Knud Rasmussen, chiefly with a view to benefit the natives of the far north which are the finest representatives of the Eskimo race in

PLATE IV

FIG. 4. ICEBERGS STRANDED IN DISKO BAY OPPOSITE
THE ARCTIC STATION

FIG. 5. THE RED RIVER FLOWING INTO DISKO BAY, NEAR
THE ARCTIC STATION

PLATE V

FIG. 6. BEACH NEAR THE ARCTIC STATION
Volcanic breccia with basaltic plateau behind

FIG. 7. THE BASALTIC PLATEAU NEAR GODHAVN
Rounded hummocks of gneiss in the middle distance ; in the foreground grass on the sandy beach

Greenland. The American traveller, Peary, who was one of the pioneers of exploration in the extreme north and first demonstrated the insularity of the country, caused the natives 'to emerge from the Palaeolithic conditions under which they had hitherto been accustomed to live.' Rasmussen, who is one of the best friends the Eskimoes ever had, determined to found a permanent Settlement for the purpose of preserving and continuing the good influence of Peary and as a base for his own expeditions. Thule is the most northerly Arctic Settlement in the world. Knud Rasmussen and Mrs Rasmussen were our fellow-passengers on the voyage from Copenhagen and from him I had an account of the objects of the Fifth Thule Expedition upon which he was about to embark. News has been received of the satisfactory progress of the expedition which is chiefly concerned with the investigation of the customs, language, and folk-tales of the groups of Eskimoes scattered over the Arctic Archipelago. In addition to Danish officials returning from leave the passengers included two Danish bishops, a church dignitary from Norway, and others who were going to take part in the Hans Egede celebrations. The Primate of Denmark, Bishop Ostenfeld, was availing himself of the favourable opportunity afforded by the Hans Egede celebrations of paying a visit to his diocese beyond the sea. On the roughest day one was always sure of finding the bishop happily smoking a pipe with a bowl of unusual capaciousness. There were also two Swedish naturalists, the Rev. Dr Enander and Dr Floderus, who are well known as authorities on Arctic Willows, also a young lady-doctor from Copenhagen, Miss E. Svensgaard, who had pluckily accepted for one year the post of Medical Officer in the Egedesminde district of North Greenland.

On August 17 I saw Knud Rasmussen's motorschooner sail from Godhavn (Disko Island) for Godthaab with members of the Fifth Thule Expedition on board, some of whom, including the leader and his companion in many expeditions, Peter Freuchen, expected to be absent at least three years. At about 2 a.m. we took leave of the party after many hours spent in most vigorous dancing in which both Danes and Greenlanders joined: the schooner slowly disappeared behind the rounded hills of gneiss that enclose the harbour of Godhavn (Fig. 9), the tall bearded figure of Peter Freuchen at the helm like that of a Viking steering his ship into the unknown.

The Danish Arctic Station at Godhavn, where we spent several weeks botanising and geologising in the neighbourhood and working in the library and laboratory, is the only station of its kind where adequate facilities are provided for experimental work within the Arctic Circle[1]. A series of valuable scientific papers containing the results of research, carried out by the Director of the

[1] A fuller account of the Danish Arctic Station was published in *Nature*, November 3, 1921.

Station and by other workers, has already been published. The building is at present too small to meet the increasing demands upon it and additions are in contemplation. Its conception and erection are due to the energy and determination of Mr Morten P. Porsild (M.Sc. of the University of Copenhagen), who, with the help of grants from private sources, equipped the very comfortable building which was taken over by the Danish Government with Mr Porsild as the first Director in 1906. It would be difficult to find a more suitable place for the training of men desirous of qualifying themselves for scientific Polar exploration or a man better fitted by knowledge and experience than the present Director to superintend not only the necessary scientific work, but also to give instruction in sledging and in other useful arts. Writers of books or scientific papers on Polar exploration or natural history render good service to science by sending copies of their publications to the 'Director of the Danish Arctic Station, Godhavn, Greenland, *via* Copenhagen.' The building, including a well-equipped laboratory, a library, and living rooms, is situated at the foot of the low hills of gneiss close to the shore of Disko Bay behind which the flat-topped hills of horizontal basaltic lava and volcanic ash rise to a height of over 2000 ft. (Frontispiece). My bedroom windows looked out across the bay to the low rocky islands in the distance which lie between Disko Island and the Greenland coast: near the beach, a few hundred yards from the house, groups of stranded icebergs lay in deep shadow or in brilliant sunshine (Fig. 4). Going to bed by day in a situation such as this had its advantages, though at first sleep came slowly. Greenland has many and substantial advantages over a country possessing all 'modern conveniences'; not the least of these is the practice followed at the Arctic Station of not being called in the morning. There are no duties to be performed requiring punctual attendance and the breakfast of coffee and bread and butter does not suffer by neglect; it is possible to enjoy the rare luxury of getting up when the desire for sleep has been satisfied. A short distance to the left the Red River flows into the bay (Fig. 5) and in certain lights a sharp boundary-line separates the turbid freshwater reddened by basaltic sediment from the clear blue sea-water beyond, an excellent demonstration of the transportation of sediment and of rock-building on the edge of a coast. Close to the Station a steep bank covered with vegetation rises from the lower marshy ground bordering the shore to the foot of the volcanic rocks above. The dark green of the bank is traversed here and there by a band of brilliant emerald green Moss (*Webera albicans*) that marks the course of a clear stream. Notices in the Eskimo language were put up some years ago by the Director asking natives to abstain from gathering plants for fuel, as the locality is very rich in interesting members of the local flora: this request has almost invariably been respected.

Shortly before we finally left the Arctic Station

early in September the vegetation had assumed the rich autumn colouring. Seen from the edge of the sea on an evening at the beginning of September the beach of olive green and brown sand was lit up with clumps of golden yellow Sea Purslane (*Arenaria peploides*); the swampy ground in the middle distance was partially covered with the crimson foliage of creeping bilberry bushes interspersed with green and orange patches, and above the steep and still green mossy slopes the escarpment of dark brown volcanic rock was sprinkled with clinging colonies of yellow vegetation. Farther in the background, with dark shadows on their steep sides, were the flat-topped mountains, huge truncated pyramids cut out of a once continuous plateau built up of successive layers of dark brown lava and redder strips of volcanic ash (Fig. 6). The photographs, (Figs. 6 and 7) taken after a recent fall of snow, show very clearly the layers of rock above the talus slopes.

Lying on the beach of Godhavn harbour is the broken and battered hull of the 'Fox,' an English yacht of 177 tons which provides an interesting link with the middle of last century (Fig. 8). In 1857, after the decision of the English Government against the despatch of further expeditions in search of Sir John Franklin, Lady Franklin purchased the 'Fox' which had been built a year previously as a private yacht: the vessel was refitted at Aberdeen and sailed for the Arctic regions under the command of Captain Leopold M'Clintock. After wintering in the pack-ice of Melville

Bay (between Cape York and lat. 75° N.) she drifted a distance of nearly 1400 miles through Baffin Bay before getting free[1]. On returning from the expedition, which had been successfully accomplished, the 'Fox' was sold to a firm at Copenhagen for sealing off Spitsbergen, and afterwards passed into English hands as a surveying ship in connexion with a projected cable between America, Greenland, and Iceland. Later the yacht was used as a cargo boat by the owners of the cryolite mine in South Greenland and by them sold to the Danish Government for coastal service. In 1912 she was placed temporarily at the disposal of a Swiss Expedition on the west coast, the members of which erected a monument—a large metal gas cylinder—in memory of the 'Fox' ('Til Minde om Fox') on the rocks near Jakobshavn (Map B, J). On returning from the expedition she ran aground but was refloated and eventually reached Godhavn, where she was condemned as unfit for further service. The Director of the Arctic Station, with his motor-boat, towed the 'Fox' to her last resting-place in 1913.

A more remarkable instance of southerly drifting of ice from North Greenland is worth recalling. An ice-floe carrying nineteen persons, including nine Eskimoes, from the ill-fated American ship 'Polaris,' which was abandoned in Smith's Sound in October, 1872, made a longer voyage than the 'Fox'; the party was picked up in April, 1873, forty

---

[1] *The Voyage of the 'Fox' in the Arctic Seas*, by Sir F. L. M'Clintock, London, 1859.

miles off the coast of Labrador. During the voyage the Eskimo hunters were able to obtain a supply of fresh food sufficient for the preservation of life. One member of the party, who was in 1872 a girl of eleven years old, is still living on Disko Island.

On the summit of a rounded headland of gneiss overlooking Disko Bay (Fig. 9), near the entrance to Godhavn harbour, is a small wooden hut, its sides made of boards nailed to four large pieces of the jaws of Greenland whales, which was formerly used as a look-out station for whales. Many names are cut in the boards of the hut, including those of explorers and of sailors from British ships dated more than a century ago. Standing by the look-out with the breakers of Disko Bay rolling up the rocks below, one thought of the members of many British Arctic expeditions who visited the spot while their ships lay at anchor in the harbour before sailing to the polar seas.

From Godhavn we made two motor-boat expeditions, travelling over 600 miles in five weeks. After a short excursion along the east coast of Disko Island in the 'Clio borealis,' the motor-boat belonging to the Arctic Station, we returned to our base, unfortunately, owing to a broken shaft and the consequent delay, a few hours too late to be present at a luncheon on board the King's ship to which we had been graciously invited. Our second and more extended excursion was made in the 'Angut,' a motor-boat belonging to the Director for Greenland, Mr Daugaard-Jensen, who generously placed it at our disposal when he heard of the disablement of the boat belonging to the Arctic Station in which we had made our first trip. We proceeded round the south and west coasts of Disko Island to Hare Island (see Map *B*), thence to Upernivik Island, across Umanak Fjord to the island of Umanak and from there we coasted along the north shore of the Nûgssuaq Peninsula, calling at many places to collect specimens, and round to the south coast which is separated from Disko Island by the Vaigat, a strait about eight to ten miles in breadth and ninety miles long. Navigation in the Vaigat is not always easy; sudden squalls are frequent; there are very few parts of the rocky coast, most of which is uninhabited, where shelter or good anchorage can be obtained, and icebergs are often inconveniently numerous (Fig. 10). After calling at Jakobshavn and other places on the mainland we returned to Godhavn, which we finally left on the 'Hans Egede' on Sept. 6. Throughout our wanderings we were everywhere received with friendliness and most willingly assisted both by Danish officials and by Greenlanders.

On disembarking from the ship's boat at the small landing-stage of a Settlement we were greeted by a happy and curious group of natives and at the larger Settlements also by some of the Danish residents, by whom we were hospitably entertained. The European residents keep open house when a ship is in harbour and a cordial welcome is given to foreign visitors. Though much of the food used by the Europeans is imported

PLATE VI

FIG. 8. GODHAVN HARBOUR
The arrow points to the hull of the 'Fox' on the beach

R. E. Holttum phot.

FIG. 9. GODHAVN, FROM ABOVE THE ARCTIC STATION
The arrow points to the outlook station

PLATE VII

FIG. 10. ICEBERGS IN THE VAIGAT

FIG. 11. JACOBSHAVN ICE-FJORD
Looking out to sea

from Denmark, the animals of the country often played a prominent part in the meals to which we were invited. Birds, such as Eider Duck, Auks, Ptarmigan, and Gulls, also eggs of the Tern, Auk, and Guillemot were often provided; sometimes we had the good fortune to obtain Reindeer meat and excellent Arctic Salmon, as well as many other kinds of fish. Seal meat, though its almost black colour makes it at first sight unattractive, is good eating; and raw Porpoise hide (matak) is by no means to be despised—it is said to combine a delicate taste of nuts and oysters: a slight acquaintance with it hardly justifies this opinion.

I cannot speak too highly of the services rendered by Mr Porsild, the Director of the Arctic Station; he made all the necessary arrangements for our motor-boat excursions, and was always ready to give us to the fullest extent the benefit of his experience acquired in the course of many years' residence in Greenland and of his unrivalled knowledge of Arctic natural history. Several of the facts recorded in this sketch were supplied by him. His son, Erling Porsild, who has a good knowledge of the Eskimo language and an intimate acquaintance with the flora, was an invaluable companion. To Mrs Porsild we were much indebted for the kindly hospitality which she provided.

CHAPTER II

Yes, as I often think, it is not the poetical imagination, but bare science that every day more and more unrolls a greater Epic than the Iliad; the history of the world, the infinitude of space and time!...it is in itself more wonderful than all the conceptions of Dante and Milton.   FITZGERALD.

Greenland a land of sunshine. Physical features and scraps of geological history. Evidence from fossil plants of warmer climates in the past. A limitless desert of ice; glaciers and icebergs.

TO most people Greenland suggests 'icy mountains,' barrenness, and barbarous natives, a conception resting upon a basis of fact, but which does scant justice to a land characterised by grandeur of scenery and by features of exceptional interest to a student of evolution in the widest sense, nor is it just to the natives whom to know is to appreciate and admire.

As early as the thirteenth century an account was written by a learned Norseman whose name is not revealed[1] in a work entitled *Konungs Skuggsja* (*The King's Mirror*), which shows that many centuries ago the more striking characteristics of Greenland were not unknown. This account is in the form of a dialogue between a father desirous of imparting information and a son whose function it is to stimulate the father by apposite questions. Fact and fancy both have a place in the volume[2].

[1] Some suggestions as to the author's name and the date of the MS will be found in Dr Nansen's book, *In Northern Mists*.

[2] The translation, from the old Norwegian, from which the following extracts are taken is by Prof. Larson of Illinois and was published in 1917 under the title *The King's Mirror*.

'It is reported,' says the father, 'that the monster called Merman is found in the seas of Greenland ...another prodigy called Mermaid has also been seen there.' In this connexion reference may be made to a description of a Merman quoted by Mr Whymper in an article on his travels in Greenland (*Alpine Journal*, vol. v, 1870) which, he states, 'clearly refers to a kayaker.' An Eskimo in a kayak, his body clothed in skins, might well produce the impression of some strange or mythical apparition.

Fancy with fact is just one fact the more.

The father adds: 'Only a small part of the land thaws out, while all the rest remains under the ice ...all the mountain ranges and all the valleys are covered with ice...the land has beautiful sunshine and is said to have rather a pleasant climate.' The son replied, as many people who have not been to Greenland would also reply, ' it is hard for me to understand how such a land can have a good climate.'

Greenland in the summer is truly a land of sunshine, and for many weeks the sunshine is continuous; the temperature in the warm season seldom falls below the freezing-point. The rainfall in the south is relatively high, about six times as much as in the north: in the district where our work lay the average annual rainfall is about nine inches. The accompanying chart, simplified from one published in the recent official book on Greenland[1], gives in a convenient form certain meteorological

[1] *Grønland* (Copenhagen, 1921), vol. I, p. 162.

data collected at Jakobshavn (lat. 69° N.): Curve I shows the air temperature in degrees Centigrade and Fahrenheit for the twelve months; II shows the percentage of days in each month on which the thermometer fell below the freezing-point; III gives the rainfall in inches and millimetres. It is interesting to note that the date of flowering

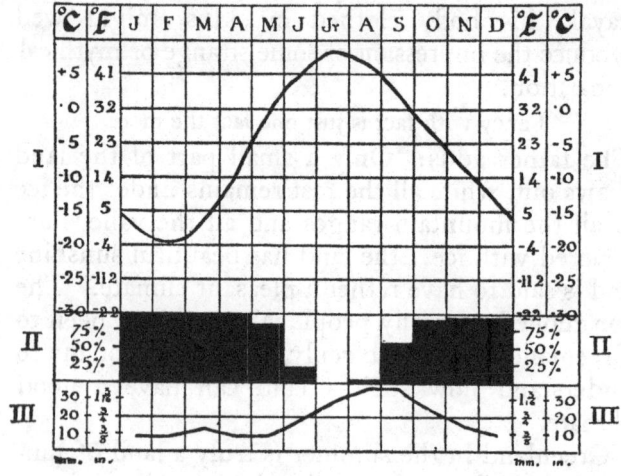

of several Greenland plants fluctuates within a range of a few weeks and that this is consistent with the variation of the date of incidence of the spring temperature, that is a temperature above freezing-point. In the chart the temperature curve (I) shows the mean about which there is a fairly wide fluctuation.

It is on the whole true that the further north one goes, the more settled and finer the weather.

After a hot climb up the steep slopes that border the sea it is a welcome change to lie on the heath enjoying the cool breeze, the wonderful panorama of mountain, sea, and ice and to revel in the peaceful solitude.

From the granitic headlands of Cape Farewell, on the latitude of the southern extremity of the Shetland Islands, Greenland extends slightly beyond lat. 83° N.; it is nearly 1700 miles long, a distance equal to that from the northern limit of the Shetland Islands to the north coast of Africa, and has an average breadth of about 600 miles, an area approximately four times that of France. On the north-west Greenland is separated from Grant Land, Grinnell Land, and Ellesmere Land by the narrow channels connecting Baffin Bay with the Polar Sea; so narrow are the channels that Eskimoes can easily pass across. It was doubtless by this route that the ancestors of the present Greenlanders reached the country. With Europe Greenland is closely connected geologically. In the remote past, at least, there was probably a vast continent, a northern Atlantis, stretching from what are now the highlands of Norway and Scotland to the Arctic regions of America. Greenland is in a geological as also in a biological sense a connecting link between the Old and the New World. By far the greater part of the island, so far as it is possible to ascertain the structure of a land almost completely covered by ice, consists of coarsely crystalline rocks mainly of igneous origin and of an antiquity that is inconceivably remote.

The cliffs on some parts of the coast are built up of limestones, sandstones, shales, and old pebble beaches containing the remains of animals and plants characteristic of several geological periods and clearly indicating climatic conditions within the Arctic Circle much more genial than those at the present day. Even in the extreme north, on the shore of the Polar Sea, limestone rocks have been described by the Danish geologist, Koch, as veritable coral reefs of the Palaeozoic era.

The most northerly point at which fossil plants have been found is on the east coast of Greenland, between lat. 80° N. and 81° N. Fragmentary remains of plants were found by the Denmark Expedition of 1906–1908: these were described by the late Professor Nathorst, who recognised them as members of a flora which preceded that of our Coal Measures. The locality where these plants were found is nearer to the North Pole than any previously recorded for Carboniferous plants.

Greenland is a mountainous plateau mainly composed of some of the oldest rocks in the world belonging to a stage in the history of the earth (the Archaean period) which is shrouded in mystery: of the life of this period we have no certain knowledge. On the extreme northern coast, also at many places on the east and west coasts, the presence of thick series of ancient sediments and of rocks consisting of accumulated masses of the calcareous skeletons of marine animals is evidence of recurrent subsidences of the land and the submergence of its edges. The occurrence of terraces of sand and gravel at a height of from two to three hundred feet above the present tide-level—several of which we examined—containing marine shells of species still living in the Arctic seas proves an upward lift of the coast-line in comparatively recent times. A still more recent movement, but in a downward direction, is demonstrated by a comparison of a series of photographs, taken over a period of several years by the Danish geologist, K. J. V. Steenstrup, which shows that the tangled mass of brown seaweed which clings to the foot of the cliffs at low water is slowly creeping upwards. The fact that iron rings for ships' cables fastened into the rocks on the west coast are exposed only at low tide is confirmatory evidence that, on the west coast at least, Greenland is sinking.

The fossil-bearing rocks it was our aim to investigate are exposed along the shore and in the ravines of Disko and other islands and especially on the Nûgssuaq Peninsula. Most of them were deposited during the Cretaceous period; others are Tertiary in age. Slabs of rock detached with the aid of a pick-axe from the side of a ravine where the hills are made of a succession of sheets of sediment—the sands and muds of some ancient lake or lagoon—are found to be covered with the clearly outlined impressions of large leaves like those of the Plane or Tulip tree, fronds of ferns hardly distinguishable from species (of the genus

Gleichenia) living to-day in tropical and sub-tropical countries; there are also twigs of Conifers, some of which are almost identical with those of the Mammoth tree (*Sequoia* (*Wellingtonia*) *gigantea* now confined to a narrow strip of the Californian coast), and massive stems of forest trees. None of the leaves preserved in the Greenland rocks have a greater fascination for the student of the past history of living plants than those of the genus *Ginkgo*. This genus is now represented by a single species, the Maidenhair tree (*Ginkgo biloba*), which is sometimes said to occur in a wild state in China, though it is probable that even in China and Japan, where it grows abundantly, it is only as a cultivated tree associated in the oriental mind with some religious symbolism. *Ginkgo* is often planted in gardens and parks in Europe and America and is distinguished from all other trees by its broad and often lobed, wedge-shaped leaves. Fossil leaves, some indistinguishable from those of the sole survivor of this ancient genus, have been found in the Cretaceous sediments on Upernivik Island (Map *B*), in sedimentary rocks associated with basaltic lavas at Sabine Island (lat. 75° N.) on the east coast of Greenland, at several localities within the Arctic Circle, also in many other regions of both the Old and the New World. These records afford an exceptionally striking illustration of the possibilities offered by a study of the herbaria of the rocks of connecting the present with the past, of following the wanderings over the world and of tracing the rise and fall in

PLATE VIII

A   B

C   D

E   F

FIG. 12. SKETCHES ILLUSTRATING TYPES OF ICEBERGS

PLATE IX

FIG. 13. ICEBERG OFF THE COAST OF DISKO ISLAND
The upper part of the iceberg consists of ice-breccia

FIG. 14. UMANAK FJORD
The bowsprit of the boat points to Umanak Island

their fortunes of still living members of the plant kingdom. These fragmentary relics, 'the ghostly language of the ancient earth,' suggest problems that are more easily stated than solved.

Many records of ancient floras are readily decipherable, foliage shoots and clearly outlined leaves showing the finest veins, the plant substance changed into a thin film of coaly substance which on treatment with certain chemicals reveals under the microscope details of the surface cells and throws light both on the affinities of the plants and on their relation to the world in which they lived. The minute structural details of petrified wood after it has been cut into transparent sections can be examined with as much thoroughness as those of a living stem; the living substance has gone, but the framework remains and through it we obtain an insight into the mechanism of the plant which was alive some millions of years ago. Other fossils are but 'age-dimmed tablets traced in doubtful writ,' and these add zest to the task of interpretation.

Two among the many problems which exercise the ingenuity of geologists and botanists may be mentioned: if, as seems certain, the climate of Greenland was warm enough to support a vegetation including forest trees and other plants closely related to species now growing in warm temperate and sub-tropical districts in North America, southern Europe, China and elsewhere, what causes can be invoked to produce the necessary change? Of the plants that exist in Greenland some occur also in different parts of Europe, others have their nearest relatives in North America: where was the original home of the Arctic floras and what was their fate during and subsequent to the Glacial period which reduced North America and north and central Europe to much the same state as that of Greenland to-day? When the Glacial period was at its maximum, Greenland was even more completely covered with ice than it is to-day and the probability is that the whole of the flora was destroyed. With perhaps one exception, a species of Pondweed (*Potamogeton groenlandicum*), the four hundred odd species of flowering plants and fern-like plants which have been recorded by botanists from different parts of the island include none that are peculiar to Greenland; they are either American and circumpolar types or southern species. The American and circumpolar plants doubtless arrived, when climatic conditions became possible for the existence of the higher plants, by way of Smith Sound on the north-west corner, while most of the immigrants from the south and west were transported across the sea by natural agencies. A few were no doubt introduced by the old Norse colonists. It has been assumed, though on insufficient evidence, that a land connexion existed in post-glacial times between East Greenland, Iceland, and the Faroes[1].

---

[1] Mr Holttum has recently written an account (*Journal of Ecology*, vol. x, No. 1, 1922) of the vegetation of West Greenland summarising the present state of knowledge from an ecological standpoint. A botanical bibliography is appended.

Thomas Hardy in *The Return of the Native* speaks of Clym Yeobright walking alone on the heath 'when the past seized upon him with its shadowy hand, and held him there to listen to its tale. His imagination would then people the spot with its ancient inhabitants.' Similarly the waifs and strays from the vegetations of the past enable us with a certain degree of accuracy to reclothe the hills with plants of other days and other climes. It is impossible with precision to interpret in degrees of temperature what the buried leaves and twigs indicate; but we may safely say that they belong to plants which could not have existed under conditions comparable to those endured by the present Arctic vegetation. One of the most convincing and impressive arguments in support of the prevalence of an almost, if not quite, tropical climate in Greenland during the Cretaceous epoch is furnished by portions of large leaves and pieces of the fruit of a Breadfruit tree discovered by members of a Swedish expedition in 1883 on the coast of Disko Island and described by the late Professor Nathorst, who was well known as an Arctic explorer and an exceptionally able student of the floras of the past. The Breadfruit, *Artocarpus incisa*, which the Greenland fossil closely resembles, is cultivated practically all over the tropics and is native in some of the Pacific Islands. The fossil *Artocarpus* was found at Ujaragsugssuk (Map *B*, U), on the shores of the Vaigat, a place at which we failed to find any good specimens of fossil plants. It is interesting to note that species of

Artocarpus, which is now exclusively an Old World genus, are recorded also from North American rocks.

Seen from the sea, the coast of Greenland forms a long line of mountains often reaching a height of 3000–4000 ft.; the darker blue of the nearer hills shading gradually up the deep and tortuous fjords into the lighter tones of those farther inland. Off many parts of the coast lie scattered groups of islands, or skerries, like huge round-backed whales, the ice-worn summits of a submerged mountain-range.

Over the whole of the interior is the 'dead storm-lashed desert of ice' rising in the central regions to a height of 8000–10,000 ft., its surface thrown into gentle undulations and the monotony occasionally broken by a stream that plunges with a roar into a chasm of unknown depth. Fridtjof Nansen, who in 1888 was the first to cross Greenland, compared the inland ice to the gently sloping surface of a shield many hundreds or even thousands of feet in thickness. As the sloping sides of the ice-shield approach the edge of the plateau crevasses are of frequent occurrence, and here and there a few of the higher peaks of the buried highlands, with groups of the more hardy Arctic plants adhering to their rocks, stand as lonely sentinels on a limitless field of snow and ice. From the inland ice glaciers, like huge tentacles, are thrust outwards towards the sea, far surpassing in the rate of movement the glaciers of the Alps, and as the ice reaches water deep enough to buoy up the moving mass portions are broken off as icebergs. The face of the Humboldt glacier, the largest of these tentacles (Map *A*, on the edge of the north-west coast, just south of lat. 80° N.), is an 'abrupt and threatening precipice' 300 ft. high and sixty miles in breadth. One of the most prolific berg-forming glaciers on the west coast stretches across the head of the Jakobshavn Ice-fjord (lat. 69° N.), a few miles to the east of the Jakobshavn Settlement. It has been calculated that the daily discharge of ice through this ice-fjord amounts to 432,000,000 cubic feet. The ice discharged annually from the Jakobshavn Ice-fjord would, it is stated, make a mountain two miles long, two miles broad, and a thousand feet high. The surface of the water, as seen from the hummocky coast behind the Settlement, is a continuous mass of ice; icebergs, some floating, some stranded, are huddled together in disorderly array suggesting the fall of a stupendous avalanche into the waters of the fjord (Fig. 11). At the western end of the fjord the icebergs set out to sea, drifting, it may be, many hundred miles before they melt or come to rest on the shore of Newfoundland, or even farther south, breaking up like ships aground. It may well be that some of the icebergs encountered during a voyage to the St Lawrence or to New York began their journey in the Jakobshavn Fjord. Others make shorter voyages and, after drifting across Disko Bay or into the Vaigat Strait, end their course in home waters. In size the Greenland icebergs vary enormously: the highest surpass those in the

Antarctic seas. Members of a German scientific expedition to the west coast measured a monster berg that was over four hundred feet above the water, but the highest usually seen probably do not exceed two hundred feet. The portion seen above water when an iceberg is floating is about one-eighth of the whole vertical thickness of the mass. Mr R. E. Priestley tells me that, on the average, Antarctic icebergs, which have more air included in the ice than Arctic bergs, and are therefore lighter, have about one-sixth of their mass above water-level.

> The glittering berg, the mariner's foe,
> Rears pinnacled peak on high,
> But few are there know how far below
> That isle of emerald ice and snow
> Its dark foundations lie.

It is a fairly common experience to see a large berg of more or less rectangular form penetrated by a tunnel, like a massive arch that is gradually broadened by the action of the sea and sun at the expense of the sides and roof, until the whole breaks into two or, by the destruction of the crown of the arch, two pinnacled bergs are produced. It has been suggested by a Swiss writer, Prof. Mercanton, that many of the innumerable forms assumed by floating icebergs are derived from tabular blocks through which water has driven a tunnel. Such observations as I was able to make lend support to this hypothesis.

The sketches of icebergs stranded in Disko Bay opposite the Danish Arctic Station, reproduced in

Figs. 12, A–F, illustrate a few forms additional to those shown in the photographs. Fig. A shows a large berg probably about 150 ft. above the water, weathered into its present shape from a tabular block detached from a glacier: Fig. A was drawn on August 29; by August 30 the collapse of the arch (Fig. B) had converted the berg into two, apparently separate bergs connected below the water-level. A similar alteration of form due to the rupture of a large central aperture is illustrated by Figs. E and F. The black patch in the lower part of the white cliff of the massive tabular berg seen in Fig. C represents an area of brilliant cobalt blue, which probably means that formerly, when the ice was still included in a glacier, there was a channel occupied by a stream: the water subsequently froze into a brilliant blue mass. A not uncommon type of surface sculpture is shown in Fig. D: unequal melting produced a series of parallel grooves and ridges.

The life-history of a large iceberg towering 100 ft. or more above the sea and with a much greater mass—varying according to the density of the ice —below the water, would make an interesting story. When calved from the face of a glacier an iceberg may be launched as a flat tabular block a few hundred feet in length; for a time it retains its original form, but as it drifts to sea and is exposed to the wash of the waves and encounters different temperatures, air at high tension, imprisoned in cavities in the ice, has the pressure reduced and this acts like an internal explosion, causing portions of the berg to burst with a loud report.

A doubtful throne is ice on summer seas.

On a still night the booming of breaking icebergs is often the only sound that disturbs the silence. If the ice is not too far away, the effect of the loss of a part of the mass as the detached pieces fall with a roar into the water is seen in the heaving of the iceberg which slowly and majestically rises and sinks, and may turn completely over, eventually coming to rest in a new position and with an entirely different aspect. The words of the Ancient Mariner,

> The ice did split with a thunder-fit...
> It cracked and growl'd, and roar'd and howl'd,

aptly describe the rending of a large berg. The smooth grooves and ridges on the side of the iceberg above the water-level seen in Fig. 13 were formed by wave-action and indicate the shifting of the equilibrium of the ice since they were produced. The upper part of the berg consists of angular pieces of ice embedded in a dirty matrix, a fairly common type of structure which may have its origin in the débris filling a large crevasse in the parent glacier.

There are few more superb sights in nature than a sheet of water with floating icebergs. The scene over Umanak Fjord (Fig. 14), with some of the highest mountains in Greenland on its northern and eastern shores, comes to my memory. It was a perfect night: a clear sky, and the sun not far above

PLATE X

FIG. 15. MOUNTAINS ON THE EAST SIDE OF UMANAK FJORD

FIG. 16. UMANAK FJORD AND UBEKJENDT ISLAND
FROM UPERNIVIK ISLAND

PLATE XI

FIG. 17. ESKIMO HOUSES ON MANÎTSOK ISLAND
The woman is standing at the entrance to a house;
above the man are two kayaks

R. E. Holttum phot.

FIG. 18. WOMEN CUTTING UP FISH, JAKOBSHAVN HARBOUR

the horizon over Baffin's Bay usurping the office of the moon, which appeared as a ghostly disc above the snow-capped basaltic hills of the Nûgs-suaq Peninsula. On the surface of the sea floated innumerable icebergs, tabular masses sometimes with an arch cut by some glacier stream and enlarged by the action of waves, bergs with pinnacles or leaning towers, others assuming the form of some gigantic bird or sea-monster. The water was smooth as glass except where falling pieces of ice, trailing slowly from the parent berg in lengthening lines of white, made advancing circles of gentle rollers. Some of the bergs reflected a rosy light; others seemed to be shining blocks of Carrara marble shading near the undercut base into a brilliant green-blue; bands of deep blue like inlaid strips of lapis lazuli, stretching across the opaque whiteness, showed where fissures had been filled with clear ice free from the included air which produces the marble-like opacity. To the west, high massive cliffs of islands or projecting headlands with jagged peaks of gneiss (Fig. 15) made a striking contrast both in form and in their glowing redness to the dark purple hills of the mainland, their flat tops crowned with low white domes of ice. To the east was one lofty peak encased in snow like a polished pyramid of marble. The view reproduced in Fig. 16 was taken at a height of a few hundred feet in a valley at the south end of Upernivik Island. The ridge of hills at the western end of Umanak Fjord is part of Ubekjendt (Unknown) Island; this island consists of basalt except the broadly conical peak with snow on its flanks, near the south end of the ridge, which owes its greater height to the presence of a core of a different kind of rock. Farther away on the left is dimly seen the coast of the Nûgssuaq Peninsula.

CHAPTER III

Our Country has wide borders...and it bears secrets in its
bosom of which no white man dreams. Up here we live two
different lives; in the summer under the torch of the warm
sun; in the winter, under the lash of the north wind. But it is
the dark and cold that make us think most.

Saying of a Greenlander quoted by KNUD
RASMUSSEN in *The People of the Polar North*.

The Eskimoes and their rulers. Greenland a 'closed' country.
Danish methods of government. Life, personal characteristics,
language, and dress of the Eskimoes. The kayak and the umyak.
Coal in Greenland.

NANSEN in *Eskimo Life*, a book which gives
an admirable description of the people
written with sympathetic appreciation, though the
condemnation of the Danish method of government is much too sweeping, thus sums up his impressions of Greenland and the Greenlanders: 'It
is a naked lonely land, like no other land inhabited
of man. But in all its naked poverty, how beautiful
it is...strange is the power which this land exer-
cises over the mind; but the race that inhabits it is
not less remarkable than the land itself.' A race
which has been able to exist entirely by virtue of
its own resourcefulness and mastery over nature
in its hardest manifestations, in former days without metal, except perhaps in a very few localities
where native iron was available, and with no wood
save logs of drifted timber washed up on the beach,
compels our admiration and respect.

The present Eskimo population is about 14,000,

and with the exception of a single colony on the south-east coast, the District of Angmagssalik (Map *A*, A; lat. 66° N.), containing a few hundred people, the inhabitants are now confined to the west coast. The geographical distribution of the Eskimo race is exceptionally wide; it extends in scattered groups from the shores of Behring Strait to Greenland, a distance equal to about half the circumference of the globe. They have wandered as far south as the Straits of Belle Isle (lat. 51° N.) and there are settlements in the Hudson Bay region and on the coast of Labrador. Authorities differ on the problem of the original home of the Eskimoes, but the view that finds most favour seems to be that they came from the interior of Canada in the region north-west of Hudson Bay. Probably Knud Rasmussen's present expedition will throw fresh light on the subject. In a comprehensive and well illustrated work on Eskimo skulls (*Crania Groenlandica*) by Fürst and Hansen, published at Copenhagen in 1915, the authors express the opinion that the crania which they examined are not those of a primitive race. Throughout this large area their culture and language show a remarkable homogeneity: there are many dialects and considerable differences between them, but authorities state that the language of the Eskimoes of western Alaska has at least as much in common with that spoken in Greenland as English has with French. It has no close affinity with any other tongue; the Eskimo language is very rich in suffixes, and a single word, which is often of enormous length, does duty for a whole sentence. The two Eskimo words given below are quoted from the well-known Danish authority, Rink. In the course of a discussion why certain birds in Greenland are particularly shy, an Eskimo by the words 'maniarneqarnertik nujorqautigingikaluarqorpat,' which may be freely translated: 'presumably indeed their shyness is not the result of their eggs being removed,' politely expressed his dissent from the view that they had become so because their nests were often robbed. The words are constructed as follows: mán (eggs)-îar (remove)-ne (the act of)-qar (get)-ner (the fact of)-tik (their), nujor (shy)-qa (become)-ut (cause)-igí (have as)-ngí (not)-kaluar (indeed)-qór (presumably)-pât (they it).

I am indebted to Mr Erling Porsild for another example of a long word which is vouched for by the Rev. H. Ostermann, the leading Danish authority on Eskimo grammar, namely: eqalugssuarniarfiliarniarniarumagaluarpunga, which means: 'I should certainly like to go to the place where sharks are caught.'

Some of the Greenlanders understand Danish and a few speak it a little, but even if a native has a knowledge of Danish, he rarely attempts to use the language because of his dread of seeming ridiculous. The language, with its wealth of consonants and long words, when spoken in sonorous tones by a venerable native preacher at a Sunday evening service which I attended at Godhavn sounded not unpleasantly; the sermon was delivered with an eloquence that was impressive despite the fact that I did not understand a single word. The men sat on one side of the church and the women on the other; all joined heartily in the many hymns that were sung. Spittoons were provided and occasionally used by the officiating clergyman.

There is, no doubt, a persistent attachment to certain old superstitions behind the apparent acceptance of the Protestant religion. Situations sometimes arise when the tact and common sense of the representatives of the church or of the Danish residents are severely tried; it is at times necessary to explain to the more intelligent natives seeming contradictions between passages in the Bible, literally rendered, and current practices. One of the many instances of the clash of deeply rooted pagan ideas with the teaching of the Church that had come under his notice was related to me by Mr Porsild. An old Eskimo woman, who had become mentally deranged, was being treated as one possessed of a devil by the people of the small Settlement, where there were no Europeans, by methods which would soon have proved disastrous. A young Dane arrived in time to save the situation so far as the woman was concerned: he bluntly told the natives that their talk about devils was all nonsense. 'But how,' was the reply, 'can that be, when we read of people possessed of devils in the Lord's own book?'

At many localities on the east coast north of Angmagssalik traces of old Settlements have been found, and it has been suggested that the east coast may have been colonised originally by Eskimoes who migrated from the west over the northern edge of the island. It is, however, the opinion of Knud Rasmussen, who is familiar with that region, that the supposed route by the north coast is impossible owing to the lack of food-animals and because of the very great physical difficulties. The east Greenlanders probably travelled from the north, down the west coast and by Cape Farewell.

Along the greater part of the east coast the land is very difficult of access because of the large quantity of ice that drifts south with the polar current, but in the neighbourhood of Angmagssalik streams from the hills tend to disperse the ice, and it is usually possible for a steamer to reach the Settlement once in the course of each summer. The only route at present available by which a traveller can go from the west to the east side of Greenland is *via* Copenhagen.

The number of Danish officials in Greenland is about fifty. In the larger Settlements, or colonies, as the Danes call them, there are from 150 to 500 people; a Danish Factor, or local governor, a doctor, a Danish, or it may be a native, pastor, and perhaps two or three other Europeans. In a typical Settlement there is a mixture of timber houses built under the supervision of a Danish carpenter, and the low stone-and-turf houses of the Eskimoes. The more primitive huts are entered through a long, low passage like a tunnel, partly above ground, which leads to the living room, often decorated with coloured prints from Denmark; a three-legged stool supports a soapstone blubber-lamp, though there is now generally a stove. Along one wall of the room runs a raised wooden platform on which the members of the family sleep together. Glass panes in wooden frames have now replaced the sheets of semi-transparent intestine which used to serve for the admission of light. Outside is a mound of partially grass-covered rubbish, the accumulation of many years of refuse. From the rubbish heaps (kitchen middens) of older Settlements many valuable records of an earlier culture have been disinterred. With a few exceptions Settlements have no roads: smooth and slippery hummocks of rock, at least in the districts where gneiss is the dominant rock, or tussocks of grass and patches of bog have to be traversed with circumspection, and unless the native soft-soled boots are worn it is difficult to avoid falls and wet feet. Pieces of seal flesh and fish are often hung out of reach of the dogs across strips of wood fastened to long upright poles; bones of whales and other animals litter the beach, and dark-red pools lie in depressions on the rocks where the women have cut up recently captured seals. On Manîtsok Island, a few miles from Egedesminde, which we visited as guests of Miss Svensgaard, the lady-doctor in charge of the District, is one of the smallest Settlements we saw. A few stone houses, each with its long tunnel-entrance, have been built on a depression between two ice-smoothed hills of gneiss close to the edge of the sea. In front of the houses strips of fish were hung out to dry on cross-bars resting on tall

PLATE XII

FIG. 19. GROUP OF ESKIMO BOYS AT GODTHAAB

FIG. 20. AN ESKIMO IN HIS KAYAK

PLATE XIII

FIG. 21. AN UNFINISHED KAYAK
The local Catechist (Jacob Olsen) and his wife behind

R. E. Holttum phot.

FIG. 22. AN UMYAK; AT NAUJAT
A few miles west of Sarkak (S , Map *B*)

posts which also serve as supports for the kayaks (Fig. 17). On the slope of a neighbouring hill were some modern graves, with a small wooden cross at the head of each, and close to these were several much older burial-places of past generations from which the crosses had long since disappeared.

The Settlement of Jakobshavn, one of the largest in Greenland, with a human population of rather more than 400 and at least as many dogs, is an important fishing centre. At most of the Settlements there were few signs of regular activity, the inhabitants, both men and women, being chiefly occupied in critical observation of the unusual visitors, but at Jakobshavn business was comparatively brisk. The women seen in Fig. 18 are engaged in cutting up a recent catch of Hellefisk (*Reinhardtius hippoglossoides*), large flat fish resembling halibut, which are packed in barrels with salt and exported to Denmark.

Over all the Settlements except that on the east coast, which is controlled direct from Denmark, is the Director for Greenland, with his headquarters at Copenhagen, who visits the country every few years: under him are two Inspectors, one for South Greenland, with his headquarters at Godthaab, and one for North Greenland, who lives at Godhavn, on Disko Island. The boundary between the two inspectorates lies a few miles north of the Arctic Circle. At the smaller Settlements there is usually an Eskimo Manager under the supervision of the Danish Factor of one of the larger Settlements, and a native catechist, who in the winter months also acts as schoolmaster. The children do not attend school in the summer. In the Church a Greenlander may rise to the highest position under the bishop in Copenhagen, whose diocese includes Greenland. At Godthaab (lat. 64° N.) there is a seminary (Fig. 1) where the more promising Greenlanders receive further education under the Danish Principal, and those who wish to be ordained spend two years in Copenhagen. Under present conditions natives cannot become Factors or Inspectors, but as members of county and parish councils, which include no Danes, they play an important part in local government.

Two monthly newspapers in the Eskimo language are printed and edited by Greenlanders, one at Godthaab and the other at Godhavn; they contain articles of general interest with some illustrations, together with local information. The editor of the Godhavn newspaper, *Avangnâmioq*, the Northlander, is also a capable barber. Both papers are circulated without charge among the people. There is also published from time to time an official report, in Eskimo, of the number of different animals killed by hunters at the various Settlements. There is undoubtedly a good deal of latent artistic talent among the natives. Mr Porsild showed me a large collection of folk-tales collected on Disko Island and at his suggestion illustrated in colour by the Greenlander by whom the tales were narrated. The same artist, also at Mr Porsild's request, had drawn excellent and very accurate landscapes of portions of the coast of Disko Fjord showing the exact spots recorded on the official maps in Eskimo place-names. It is often very difficult on an uninhabited coast, as we frequently found, to locate with certainty localities designated by Eskimo names, many of which have reference to some not very obvious physical feature.

The small chubby boys seen in the photograph taken at Godthaab (Fig. 19) are good samples of the rising generation which, judging by the relatively large proportion of children we saw at the Settlements, is not likely to fall behind the present generation either in number or physique. Behind the group is the recently appointed Danish pastor, Mr Bugge, who was one of our fellow-passengers on the voyage from Copenhagen. The light-coloured jacket worn by the boys and by Mr Bugge is the anorak, a most comfortable and practical garment made of linen or cotton, which is slipped over the head like a sweater and is provided with a hood at the back which can be easily pulled over the head.

It is impossible within the limits of a short sketch to describe adequately the system of government: the central idea is the protection and welfare of the native population. Money made by the sale of the country's produce is spent on its administration: even in good years the margin of profit is small. From the local stores—each Settlement has an official store—articles of food and clothing can be obtained at very low prices, also tobacco, but no alcohol. So far as the native is concerned

Greenland is a 'dry' country, but the officials are allowed to have in their possession a definite quantity of alcohol; the higher the official the larger the allowance. There are no taxes in the ordinary sense, though, when skins and other things are bought by the officials from the Greenlanders, a reduction is made and the money thus obtained is spent for the country's benefit. The profits accruing to the firm which works the cryolite mine in South Greenland are taxed by the government, and this tax plays an important part in the balance-sheet of the Royal Greenland Administration. A special paper currency is used in Greenland; a 25-öre note has on it a picture of an Eider Duck, a Saddle-backed Seal is represented on a 50-öre note, a Reindeer on a one-krone note, while a Polar Bear represents five kronen. There are no police, and serious crimes are very rare. Crime is, however, not unknown nor are the criminals without a sense of humour: an *Attaché* of the Danish Legation in describing a film illustrating the royal visit to Greenland told the following story: 'Before the coast was sighted, a frail native canoe carrying a single man was detected among the ice-floes. Assuming that this was a messenger who had ventured so far from land to bid the King welcome, he was invited on board the cruiser, treated to a royal cigar and presented with a rifle. When the King landed a few hours later His Majesty was informed that the guest he had so signally honoured was an escaped convict.' Disputes between natives are settled by a court composed exclusively of Greenlanders: if the quarrel is between a Dane and a native both nationalities are repre-sented, the Greenlanders having a majority of one. The Inspector of the district is president of the court.

Whether one agrees or not with the method of government, it is at least certain that under the fatherly rule of Denmark the population is in-creasing. A Report has recently been issued by a Commission composed partly of Danes and partly of Greenlanders, which was appointed to consider the administration of the country, and it is probable that this will result in certain modifica-tions of the existing system. The continuance of the present government monopoly does not com-mend itself to all who have a first-hand knowledge of Greenland, and some of the more enlightened natives have ambitions, both natural and com-mendable, which cannot be realised under the existing régime. A native pastor whom we met, occupying a position in the Church similar to that of an English archdeacon, has given expression to the aspirations of the natives in a novel entitled *The Greenlanders' Dream*, in which the conditions in an imaginary Greenland of the future are de-scribed: this was published in Eskimo and trans-lated into Danish.

The Eskimoes, or Greenlanders as they prefer to be called, with whom we came in contact were for the most part good-tempered and cheerful and often intelligent and quick. Among the groups of natives seen at the different Settlements there were no audible signs of quarrelling: the Greenlander appreciates peace; his sharpest weapon, as a Dane familiar with the people said to me, is irony. The majority show, in a greater or less degree, signs of admixture of Eskimo and European blood. Many have Mongolian features; some resemble North American Indians, and others might pass for Europeans. A few of the men we met had thick black curly hair in contrast to the usual straight black hair. The natives that are true to type have broad oblong faces with chubby cheeks, high cheek-bones and a pointed crown, flat noses, and small dark eyes.

The men are taught at an early age the art of hunting seals, walrus, and other animals, and this involves, as the first step towards efficiency in procuring the necessaries of life, the mastery of the kayak. A Greenlander is inseparable from his kayak, the long, narrow boat (Fig. 20) which Sir Clements Markham has aptly described as 'the most perfect application of art and ingenuity to the pursuit of necessaries of life within the Arctic Circle.' The kayak and harpoon, says Nansen, 'rank as the highest achievement of the Eskimo mind.' A kayak is about 17 ft. long, rather less than 2 ft. broad and with a depth in the middle of about 9 inches. The open wooden framework (Fig. 21) is covered, except in the centre where a circular hole is left to fit the kayaker, with seal-skins that are put on in a raw or wet state, and contract on drying. The harpoon-shaft with throwing-stick and a line attached at one end to the barbed

PLATE XIV

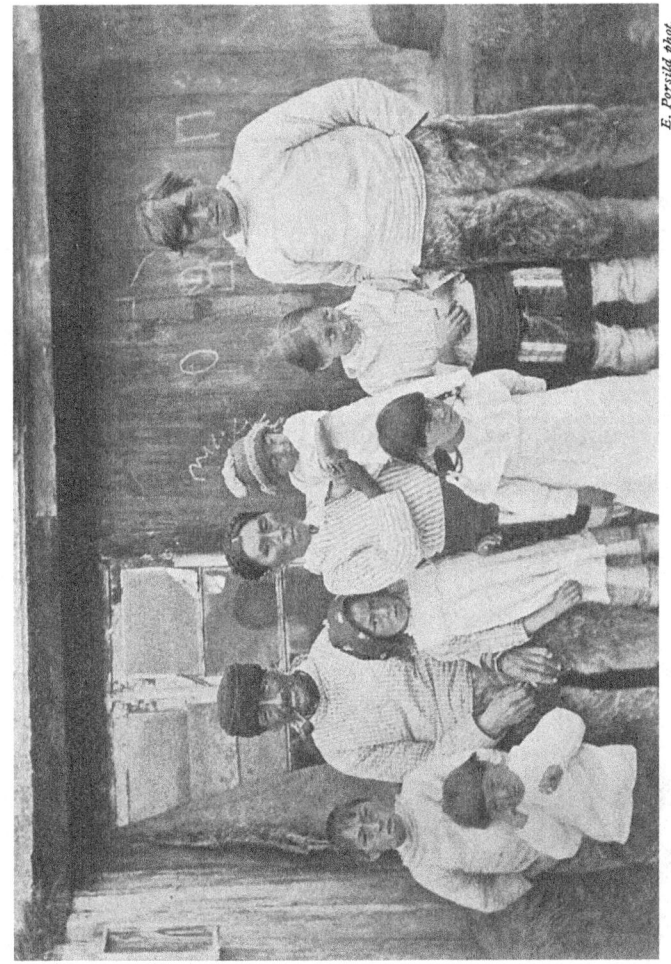

E. Porsild phot.

FIG. 23. LUDWIG GEISLER AND FAMILY; UJARAGSUGSSUK (U, MAP B)

PLATE XV

FIG. 24. THE VAIGAT AND NÛGSSUAQ PENINSULA
The 'Clio borealis' (two masts) and the relief boat

R. E. Holttum phot.

FIG. 25. THE ATA DELTA WITH COTTON GRASS (*ERIOPHORUM
SCHEUCHERZI*) AND OTHER PLANTS

harpoon and at the other to an inflated bladder, the lance, and other weapons and instruments, to which in modern times the rifle has been added, are all within easy reach of the kayaker and are held in place by pieces of hide stretched across the deck. The white linen at the bow of the kayak shown in Fig. 20 is believed to be mistaken by seals for a block of ice. A small flat piece of wood suspended underneath near the bow serves as a drift-rudder and helps to keep the boat straight when paddled against a strong wind or when it is forced backwards by the recoil of the kayaker's rifle. Narrow pieces of the bones of a whale are attached to the keel at both bow and stern as protection against ice or the rough ground on a beach. At the butt of the harpoon-shaft is a bone knob or two bone 'wings': bone is used for the fore-shaft of both harpoon and lance, also for the barbs of the bird-dart as well as for buttons and other small accessories. The fore-shaft of the harpoon and lance is held by a very ingenious arrangement of thongs of hide to the main shaft, which keeps it in place and at the same time allows the fore-shaft to be loosened and detached without being broken as it yields to the sudden jerk given by the diving harpooned seal. Walrus and Narwhal tusks, as well as whale-bone and in recent years Reindeer antlers, supply the bone used in the construction of certain parts of the kayak and its accessories. The two rather narrow blades of the paddle have strips of bone on the edges and at the ends, the chief use of which is to prevent the paddle being broken when a kayaker has to break through thin ice at every stroke as he propels the kayak over the surface of the ice. A waterproof covering of skin fits on the rim of the circular aperture in the middle of the kayak and is held in position over the shoulders of the kayaker by strips of hide fastened with bone fittings; in rough weather a complete covering is worn and many kayakers are able, if capsized, to turn completely round under water and come up on the other side without being more than superficially wetted.

There are certain peculiarities in the construction and ornamentation of both kayaks and harpoons characteristic of different localities: Mr Porsild told me that Peter Freuchen, the Danish Manager of the Thule Station in the far north and the companion of Rasmussen on many journeys, had been able to follow the wanderings of the Walrus; he found that the Walrus travels from Cape York (Map *A*, Y) to the south of Greenland along the west coast and returns to the north up the opposite side of Davis Strait and Baffin Bay. An important piece of evidence was furnished by the discovery in a Walrus taken in the extreme north of a southern type of harpoon which was eventually traced to its owner in South Greenland.

The introduction of the rifle has not been an unmixed blessing: many birds and other animals are wounded without being killed and there is much indiscriminate slaughter. In Greenland, except for foxes (May to October), there is unfortunately no 'close season.'

The fact that the kayak with its weapons and fittings is still made with a poor equipment of tools—in the past the appliances were much more inadequate—compels admiration of the resourcefulness and technical skill of the Eskimoes. It is not merely that the kayak is a remarkably efficient boat with attractive lines, but the finish of the bone fittings and the simple decorative features associated with those that are essential are evidence of artistic feeling, pride of work, and ownership on the part of the craftsman.

Another type of boat is the umyak (Fig. 22), often spoken of as the women's boat because, formerly at least, the crew, with the exception of the helmsman, who uses a scull as a rudder, consisted of women. The umyak, like the kayak, is an ancient type; boats of similar construction are used by the Eskimoes on the shores of Hudson Bay and by the Alaskan Eskimoes. A Greenland umyak is about thirty or forty feet long; it is a flat-bottomed rowing boat with a wooden framework lashed with strips of hide and bound together on the contraction of the skins stretched when wet over the sides and bottom. With the wind dead astern a small sail is sometimes used in the bow. The umyak is employed for long journeys; it can carry a considerable amount of luggage, and is easily taken on shore in bad weather when it may be used as a shelter in lieu of a tent by being inverted and with one side propped up with wooden supports. During our residence at the Arctic Station a family travelled a distance of about sixty miles in their umyak to offer a birthday greeting to the Director of the Station, and, as there were few rowers, dogs were employed to tow the boat.

On one occasion we had an interesting and exciting demonstration of the sea-worthiness of the kayak and of the courage and skill of a Greenlander. Our motor-boat, with a broken shaft, was under sail off the rocky coast of the north of Disko Island: there had been a strong breeze all day; we attempted to reach a place where there was a good harbour, but as the wind shifted we steered towards a Settlement on the mainland which on nearer view seemed to be blocked by ice, or at least there was too much ice for the safe navigation of a boat under sail. We then returned to our starting-point, the small Settlement of Ujaragsugssuk, where there is very little shelter. The wind had meanwhile increased to a hurricane. As we tacked about off the rocks, large pieces of turf were torn off the roofs of the native houses and carried up in a spiral before being flung out to sea. Our two anchors were powerless to hold the boat and there was a danger of the off-shore wind carrying us across the Vaigat on to the rocks of the uninhabited shore on the other side. It was impossible to use our small dinghy, and attempts were made by the excited natives on the beach to get a rope to us by means of inflated seal-skins, but these were blown wide of the mark. Eventually the local catechist (who is seen in the photograph (Fig. 21) standing with his wife behind a half-completed kayak which he was building) launched a kayak

PLATE XVI

R. E. Holttum phot.

FIG. 27   ENGLISHMAN'S HARBOUR

Mr Persild on the right; his youngest son, on the left

R. E. Holttum phot.

FIG. 26.   *PYROLA GRANDIFLORA*

PLATE XVII

R. E. Holttum phot.

FIG. 28. CLUMPS OF *ARNICA* AND *OXYRIA* (ON THE RIGHT)
ON THE ATA DELTA

R. E. Holttum phot.

FIG. 29. WILLOWS WITH EXPOSED ROOT ON SANDY SLOPE
ABOVE SKANSEN

and at considerable personal risk succeeded in bringing us a rope. It was a courageous act performed with apparent ease and with admirable coolness and skill. The catechist, Jacob Olsen, afterwards accompanied Knud Rasmussen as a member of the Fifth Thule Expedition.

As contrary winds prevented our return to Godhavn under sail, we despatched two kayakers as soon as the storm abated with a letter to the Arctic Station, a distance of seventy miles, and waited for the arrival of a relief boat to take us in tow. Two nights were spent in the house of the native Manager of the Settlement, Ludwig Geisler (Fig. 23), an exceptionally intelligent man skilled in hunting and in all the arts that a Greenlander practises: he is seen in the photograph with his family. In the room adjoining ours the whole family slept on the raised platform which takes the place of beds: the almost incessant coughing of the young baby served to emphasise the drawbacks of the persistent and unhygienic custom of members of Eskimo households sleeping together. We were well entertained by our host, who knew a few words of English picked up by intercourse with whalers. The drinking-water, as in many other Settlements, was obtained by melting blocks of ice, pieces of icebergs washed ashore, and stored in a large tub which gave it a slight flavour of seal oil. As it was our intention to make an early start in the morning our host wound up an alarum clock which we afterwards discovered was a most effective instrument and a refinement of civilisation not previously met with. We were awakened by a tune played in slow time and repeated many times 'diminuendo' until further sleep was impossible.

The scene reproduced in Fig. 24 is in striking contrast to that at the same locality a few days previously when our boat was at the mercy of the gale. Across the Vaigat is the mountainous coast of the Nûgssuaq Peninsula: the successive flat summits on the left exhibit the characteristic form of the volcanic rocks which cover the underlying sedimentary strata, distinguished by their yellow or bright red colour, even at a distance of ten miles or more, from the brown lavas above. The motor-boat on the left had just arrived from Godhavn, whence it had been sent by Mr Porsild who feared that the 'Clio borealis' (the motor-boat in which we were travelling) had met with a mishap. As we left Godhavn, our friend Ludwig Geisler gave us a salute of two rifle shots; a few hours later we met the two kayakers whom we had sent with a request for help.

It is the duty of Eskimo women to skin and cut up the seals, and this is performed with wonderful dexterity by the aid of a simple knife with a semi-circular blade attached to a broad wooden handle. It is said that a woman dressed in her best can cut up a seal without receiving a single splash of blood.

The chief recreation is dancing. I recall one evening at Holsteinsborg on the mainland coast when we danced in the open until midnight to the accompaniment of a concertina played with great skill by a Greenlander.

The women, like the men, wear hairy seal-skin trousers but the women adorn theirs with a band of white or coloured skin on the front of each leg: the boots, made of seal-skin from which the fur has been removed, reach to the knees; they are usually white or bright scarlet and decorated with some geometrical pattern made by sewing strips and small pieces of coloured skin to the front and top of the boots. The native boots (Fig. 39), known as kammiker, are double; into the outer covering of seal-skin fits an inner boot of dog-skin with the hair next the leg; some dried grass is placed between the two soles. The outer sole, made of the skin of the Greenland seal, is turned up all round the edge and very skilfully stitched to the upper part of the boot, made of the skin of another kind of seal (*Phoca hispida*). The skin of the sole is first chewed by the women to soften it before being stitched with sinews of whale or reindeer. The kammiker are very comfortable and warm; their soft soles, which are kept in good condition by being frequently drawn backwards and forwards over a metal edge, are admirably adapted for walking over smooth, slippery rocks, and they are water-tight. Above the trousers the women wear a broad belt and a blouse both made of some brightly coloured imported material, and on special occasions an elaborate home-made collar of open beadwork over the shoulders. The once prevalent custom among the women of gathering up the hair into a short column on the top of the head is dying out, but it is still seen, especially in the smaller Settlements. A red band round the top-knot is the badge of a maid; blue denotes a married woman; black a widow, and green an unmarried woman who is a mother.

For fuel the natives use some of the plants that grow near the Settlements, and piles of these are often seen dotted over the hill-sides to dry in the sun. Coal in some districts is abundant and easily obtained: thin seams were often exposed among the beds of shale and sandstone by the shore or in the valleys where we collected fossils. Some of the seams are of fairly good quality. The coal is available for anyone who takes the trouble to get it. There is at present one government coal-mine on the north coast of the Nûgssuaq Peninsula (Map B, C. M.) from which local schooners and occasionally the Copenhagen ships obtain supplies. A visit to this coal-mine under the guidance of the Swedish manager was an interesting experience. The coal, which crops out on the face of a cliff by the beach, is worked from slightly inclined galleries penetrating a few hundred yards inland: a short distance from the entrance all the rock-faces, including the coal itself, are covered with glistening white crystals of ice. The temperature is a few degrees below the freezing-point all the year round. In the winter when the Vaigat is frozen the coal is carried on sledges drawn by dogs. The dogs belonging to the coal-mine were some of the largest among the many hundreds we saw in Greenland. The abundance of dogs at the Settlements in North Greenland, that is, all the west

PLATE XVIII

R. E. Holttum phot.

FIG. 30. *ARCHANGELICA* AND *ALCHEMILLA*; ENGLISHMAN'S HARBOUR

PLATE XIX

R. F. Holttum phot.

FIG. 31. ORCHIDS, *POLYGONUM*, DANDELIONS, FERNS, AND OTHER PLANTS; ENGLISHMAN'S HARBOUR

coast from a few miles north of the Arctic Circle, is explained by the fact that when the freezing of the sea puts an end to the employment of the kayak, sledging is the only method of travelling and transport. Each sledge is drawn by a team of dogs, generally eight to twelve. The dogs are harnessed together and spread out in a fan-shaped group in front of the sledge and not in Indian file.

CHAPTER IV

The variety and beauty of the flora of Greenland. Some salient features of Arctic vegetation. Arctic and tropical vegetation contrasted. The Arctic and the Antarctic. Lichens and colour in nature.

A VISIT to Greenland in the summer affords a very incomplete idea of a country which is usually associated in one's mind with its winter aspect when, except in the more southern districts, the kayak is replaced by the sledge and all communication with the outer world is suspended. The isolation has compensations. A Danish friend who passes most winters in Greenland told me that he watches the last ship leave in September with a sense of relief; it means at least six months of peace and quiet. A few brief descriptions of typical scenes may serve to dispel the popular fallacy that even in the summer this Arctic land offers few attractions as a place of residence. John Davis in the latter part of the sixteenth century described Greenland as a land of desolation, and added: 'The irksome noise of the ice and the loathsome view of the shore bred strange conceits among us.' Shelley's lines,

> From the most gloomy glens
> Of Greenland's sunless clime,

though applicable to certain localities in the winter, do scant justice to Greenland in summer. It is a land with many advantages: there are no letters, no telegrams, and no public telephones—the chief Danish official at one of the larger Settlements told me with pride that he had a local telephone. Though in some districts mosquitoes may be troublesome, the fauna of the country is fortunately poor in insect pests; there are no fleas, except, I believe, in one restricted area on the west coast; no ants, and no myriapods. There are no reptiles, frogs, or rats.

The abundance of flowers makes an unexpected impression upon a visitor imbued with the idea of a country practically buried under a mass of ice of unknown depth and of a long winter when the sea is frozen and even the coastal regions are covered with snow. One effect of Arctic conditions is to limit the production of foliage shoots and often to induce an abnormal development of subterranean stems and roots and a prolific crop of flowers. The amount of energy expended in the production of roots becomes apparent if an attempt is made to dig up intact a fairly large prostrate Willow. The rocky ground is generally covered with a thin layer of soil and roots are unable to grow far in a vertical direction. In some places permanently frozen ground is met with at about two feet below the surface, while in other situations there may be at least two yards of unfrozen earth or sand in the summer. The root of one Willow we dug up was traced for at least twelve yards growing horizontally not many inches under-ground. Size is a misleading criterion of age: the wood of a Willow stem barely an inch in diameter may show as many as 100 attenuated annual rings. In the districts we visited, Willows, including the British species, *Salix herbacea* (the smallest tree in the British Isles), and a few other species with many hybrids, and the Dwarf Birch are the only trees. The tallest examples growing in sheltered places or against the sides of rocks reached a height of two to three feet; for the most part they lie prone on the ground with no main stem, but with spreading and often twisted shoots in which the annual increase in length is very small. They have water enough and in the summer abundant 'comfort of the sun,' but they are unable with impunity to grow far above the ground-level. Arctic plants point the moral of the wise words of a seventeenth century divine. 'The best means to preserve peace is in humbleness. The tall cedars feel the fury of the tempest which blows over the humble shrubs in the low valley.'

In South Greenland, on the other hand, trees are more abundant and, though usually much lower, in rare instances they reach a height of about eighteen feet. In addition to Willows and Birches there are Junipers, Alders, and the American Sorb (*Sorbus americana*).

Landing on a beach where glacial streams have built up a fan-shaped delta sloping seawards in a graceful curve from the mouth of a ravine cut by successive spring floods through the rocks of the

PLATE XX

FIG. 32. NIAKORNAT

The block on the left shows the structure characteristic of breccia

FIG. 33. THE BEACH AT GODTHAAB

Mr Holttum on the right holding one end of a large Brown Seaweed

PLATE XXI

FIG. 34. NEAR PATOOT ON THE SOUTH COAST OF THE VAIGAT
Delta, sedimentary rocks and the ice-covered basaltic plateau

FIG. 35. VIEW NEAR SKANSEN (DISKO ISLAND)
In the valley sandstone rocks capped by basaltic dykes

raised plateau, one finds stretches of muddy flats and boggy ground covered with the waving white plumes of the Cotton grass (*Eriophorum*, Fig. 25), and many other familiar plants; on the drier ground are bright reddish purple patches of a handsome Willowherb closely allied to our common British species, clumps of light yellow Poppies, and darker and more brilliant Dandelions. In both wet and dry situations the bright green feathery stems of the common Horsetail flourish in quantity. The hill-sides are often covered with a thick growth of heath-forming vegetation mixed with stumpy Willows; the leaves of some species of Willow are covered with silvery down and form an attractive background to the numberless catkins conspicuous by their bright red colour. Trailing branches of the Dwarf Birch, with smaller and less obvious flowering shoots, a parti-coloured tangle of Lichens, Mosses in different shades of green, and creeping or erect Club-Mosses (species of *Lycopodium*) are characteristic elements in the vegetation. Among the common heath plants are the Bilberry, which in the latter part of the summer provides an abundance of fruits dusted with a blue-grey bloom, the Crowberry, a *Rhododendron* resembling the Alpine Rose, a species of *Ledum*, sometimes called *Labrador tea*, a plant of American origin with dense and fragrant clusters of star-like flowers, *Phyllodoce coerulea*, its purple bells recalling those of our heaths, and an abundance of the beautiful white flowers of *Diapensia*, a genus with a wide distribution from Spitsbergen through

Grinnell Land to eastern Canada and the United States and Japan, two species of the widely spread American and Siberian genus, *Cassiope*, the commonest of which, *Cassiope tetragona*, has small crowded leaves like green overlapping scales grasping the slender stems in four regular geometrical rows with here and there a yellowish white bell pendulous on a delicate stalk.

The geographical distribution of many of the Greenland flowering plants presents interesting problems in relation not only to the efficiency of the plants as travellers, but to the position of the places from which they originally spread. Two examples may be briefly considered in this connexion.

Phyllodoce coerulea has a wide distribution: it occurs on both coasts of Greenland, in Iceland, from northern to southern Scandinavia and the mountains of Scotland, in eastern Asia, where it extends, in Japan, south of lat. 40° N.; it flourishes from Labrador to the White Mountains on the eastern side of the United States; it is found also in Alaska and in the north-west district of the Pacific coast of North America. *Phyllodoce* may have begun its existence in the northern Pacific region, probably in the latter part of the Tertiary epoch, long before the Glacial period. As the climatic conditions in the north became more severe it gradually migrated to the south, where it is still represented in the central Pyrenees and in Japan. As conditions ameliorated, the species returned to the north or ascended to the higher regions in more southern latitudes. It is an interesting fact that *Phyllodoce* does not occur on the Swiss Alps.

Cassiope tetragona is a dwarf Arctic shrub of an exceptionally wide high Arctic range; it occurs on the north-west coast of Greenland and extends a considerable distance along the west coast, but does not reach Cape Farewell; it is also met with along a strip of the northern part of the east coast. This species is widely distributed in Spitsbergen; it is unknown in Iceland, but grows on the northern Scandinavian mountains, and extends along the whole of northern Siberia, crossing the Behring Strait to Alaska, continuing thence along the northern boundary of the American Archipelago, spreading to the south-east along the more northern ranges of the Rocky Mountains to Hudson Bay, and for a considerable distance along the east coast of the North American continent and Baffin Bay to the eastern regions of the United States. Professor Rikli[1], who has published a very useful map of the geographical distribution of this and other species of *Cassiope*, believes that *Cassiope tetragona* began its wanderings in east central Asia.

[1] The following papers by Dr Rikli of Zürich contain much useful information on the distribution of Arctic flowering plants: Über *Cassiope tetragona* (L.) D. Don., *Engler's Botanische Jahrbücher*, Bd. L (Supplement), p. 268, 1914; Die den 80° N. erreichenden oder überschreitenden Gefässpflanzen, *Vierteljahrsschrift der naturforschenden Gesellschaft in Zürich*, LXII, 169, 1917; Die Arktisch-subarktischen Arten der Gattung *Phyllodoce*, *Vierteljahrsschrift* (*Zürich*), LXVI, 324, 1921. Other references to literature are given in Mr Holttum's paper quoted in the footnote on page 30.

One of the most attractive and abundant plants is *Pyrola grandiflora* (Fig. 26), a species unknown in Britain, but represented in our flora by its near relative the Winter Green: from a rosette of glossy, dark brown leaves the flowering shoot stands erect, bearing a series of wide-open flowers with pinkish white petals. The yellow and light pink flowers of species of *Pedicularis*, the genus which includes the Lousewort, crowded on stout stems with rich brown leaves add to the variety of colour. A species of *Dryas*, *Dryas integrifolia*, very similar to the British Alpine species, *Dryas octopetala*, is exceedingly common. The pure white flowers and slender grey-green stems of the Alpine *Cerastium* (the Alpine Mouse-eared Chickweed), the viviparous *Polygonum*, its tall spikes with terminal clusters of small white flowers overtopping most of its neighbours, groups of blue Harebells, and on sandy beaches the darker sky-blue flowers of *Mertensia*, several different kinds of Saxifrage, species with shining white flowers on long stalks, and the more compact cushions of *Saxifraga oppositifolia*, a species common in the Highlands of Scotland and in Arctic lands and reaching an altitude of 17,000 feet in the Himalayas, with a rich display of purple-blue flowers; species of *Ranunculus* and *Potentilla*, and an attractive little *Draba* allied to the white Vernal Whitlow grass with yellow and white flowers; clumps of yellow Dandelions and Arnicas (Fig. 28): these with many other less showy plants in which brown is the dominant shade all have a share in the general scheme of colour. Many of the Greenland flowers are familiar British or European species; others come from the New World: botanically as well as geologically Greenland has many features in common with both the eastern and western hemispheres. It is a noteworthy fact that among the flowering plants recorded from the country as a whole, about four hundred, only one or two are peculiar to Greenland. On rocky slopes, often tucked away in crevices, the cushions of the Moss Campion (*Silene acaulis*), anchored by a strong tap root like an elongated rat's tail burrowing far into the covering of earth, represent a well-known architectural type in Alpine and Arctic countries.

In the neighbourhood of Godhavn, especially in the exceptionally favourable locality known as Englishman's Harbour, so called because an English captain mistook it for the main harbour and wrecked his ship there, the abundance of southern types is a striking feature. By southern types in this sense is meant plants which reach their northern limit on the mainland considerably south of the latitude of Disko Island. The sheltered bay faces south and has the added advantage given by the warm springs, reminiscent of the days of volcanic activity in this part of Greenland, which issue along the irregular boundary between the old gneissic foundation rocks and the much more modern superstructure of basalt and beds of ash. The Willows, often assuming an espalier form against the boulders of gneiss on the slope overlooking the bay, are unusually well grown (Fig.

27); the sea Mertensia (*Mertensia maritima*), with its purple-blue flowers and glaucous leaves, is scattered among the shingle on the beach. Here can be seen in profusion, in company with a host of other plants, yard-high stems of *Archangelica* clasped by the large and handsome leaves and bearing candelabra-like umbels of small yellow-green flowers (left-hand upper corner, Fig. 30), a plant familiar to us from its use as a sweetmeat and highly prized by the Eskimo as an article of food; also the large and almost circular bright green leaves, four inches or more in breadth, and inconspicuous flowers of a northern species, which occurs in Switzerland and the Pyrenees, *Alchemilla glomerulans*, closely related to our Lady's Mantle; the tall flowering spikes of the Orchid, *Habenaria albida* (Fig. 31), akin to the Frog Orchis of Britain; also *Habenaria hyperborea* and smaller plants of the Tway Blade Orchid (*Listera cordata*), and here and there a few of the delicate and singularly attractive mauve, tasselled flowers of an Alpine Meadow Rue (*Thalictrum alpinum*):

> So blooms this lovely plant, nor dreads
> Her annual funeral.

The Butterwort (*Pinguicula*) was found in full bloom on the boggy ground. A few ferns mix their graceful fronds with the foliage of the flowering plants, and other, generally smaller ferns, pass their life hanging on the vertical faces or in fissures of rocks. The ferns include *Aspidium Lonchitis*, the Holly Shield fern, *Dryopteris linneana*, and *Cystopteris fragilis*, the Brittle Bladder fern. A

PLATE XXII

FIG. 37. PART OF THE DYKE SEEN IN FIG. 36

Mr Holttum's hammer rests on a block of sand-stone hardened by contact with the dyke

FIG. 36. A VALLEY BEHIND SKANSEN

Showing a dyke on the edge of a weathered slope of sandstone in line with the arrow

PLATE XXIII

FIG. 38. LOOKING ACROSS UMANAK FJORD FROM THE SOUTH
SHORE OF UPERNIVIK ISLAND

FIG. 39. ESKIMO WOMEN IN FRONT OF A ROOFLESS HOUSE,
UPERNIVIK ISLAND

species of *Botrychium*, the genus to which our Moonwort belongs, has also been found at Englishman's Harbour.

In the winter practically the whole of the ground is covered with snow: Mr Porsild describes the occurrence of chimney-like holes in the snow through which rises warm and damp air smelling like a greenhouse. The air temperature may be $-30°$ C., while the temperature of small pools of water among the vegetation under the snow may be as high as $17°$ C. The occurrence at Englishman's Harbour and at other localities on Disko Island of plants characteristic of the more southern parts of Greenland is consistent with an Eskimo legend, according to which Disko Island once lay much farther south. In its original home the island was a hindrance to navigation and an Eskimo sorcerer towed it behind his kayak to its present situation.

Despite the shortness of the season and the hard conditions inseparable from an Arctic climate, the vegetation competes successfully, in the show it makes, with that of warmer countries and is in some respects superior. How, it may be asked, does the vegetation of Greenland compare with that of the Tropics? Sunlight, air, and water are everywhere the driving forces of the living plant. In Arctic lands cold and dry winds and winter snow set limits to the upward growth of shoots and compel them to hug the ground and to exercise a strict economy in the production of vertical stems. A large proportion of the energy available is expended upon the formation of reproductive organs. Tropical conditions induce length of stem and leaves on a lavish scale, and the formation of dense jungles in which the competing trees make every effort to obtain a place in the sun. By comparison with the variegated carpet of flowers that brightens an Arctic landscape the ground in a tropical forest is intensely gloomy; the flowering shoots of climbers are festooned over the branches of crowded trees often blossoming far above the reach of man or even beyond his vision, while the smaller plants pass their life attached to the sunlit boughs of supporting trees in the topmost region of the jungle. Arctic conditions demand a concentration of effort and the result is a 'rush of flowers' when once the winter is passed.

> The brightest hour of unborn Spring,
> Through the winter wandering,
>
> . . . . .
>
> It kissed the forehead of the Earth,
> And smiled upon the silent sea,
> And bade the frozen streams be free,
> And waked to music all their fountains,
> And breathed upon the frozen mountains,
> And like a prophetess in May
> Strewed flowers upon the barren way.

Timely preparation is made during the growing season which ensures a prompt response to the first call of spring; buds are ready by the end of summer; in the winter they find shelter under the snow or below a covering of dead leaves. In Greenland more than in temperate regions the effect of spring is 'the miracle of earth re-clad.' It is an interesting fact that annuals are very rare; only four or five flowering plants complete their life-cycle in one season. In the Swiss Alps the percentage of annuals falls as higher altitudes are reached.

While it is true that many of the Greenland plants exhibit a characteristic and peculiar habit of growth as well as certain external characters and structural features in their foliage and stems that are usually considered to be adaptations to rigorous climatic conditions, others are in no visible respect different from plants that flourish in a warmer and much more favourable environment. The power to endure hardship probably resides in some quality of constitution, something that is fundamental in the composition of their 'physical basis of life,' the living protoplasm.

The high northern distribution and the abundance of flowering plants in Arctic regions afford a striking contrast to the conditions in corresponding latitudes in the southern hemisphere. The North Pole is surrounded by the Polar Sea bounded by a ring of circumpolar lands; the South Pole is situated on a vast continent separated from the nearest land masses by the turbulent southern ocean with scattered archipelagoes and solitary islands, some of which are of comparatively recent origin while others may be vestiges of submerged connecting bridges. Not a single flowering plant has been discovered within the Antarctic Circle. The most southerly representative of the flowering plants, over four hundred of which occur in Greenland, is a grass (*Deschampia antarctica*) which was found in the sub-Antarctic region and reaches its southern limit in lat. 62° S., a position corresponding to that of the Faroe Islands and the south of Finland in the northern hemisphere.

The fringe of Greenland where the snow and ice are discarded, like winter clothes, as soon as the freezing-point is passed, in the more favoured situations becomes a paradise of flowers not equal in brilliance to Alpine meadows at their best but characterised by a harmony of colour in keeping with the sombre grandeur of the setting. The barrenness of wind-swept slopes which on the melting of the snow are scarred by destroying streams leaving in their track patches of withered shoots pressed against the ground and dead dishevelled Willows anchored by bared roots, like cables dragged taut by the strain of rushing water (Fig. 29), intensifies the impression of sharp contrasts that a Greenland landscape produces. Charles Lamb's contemptuous description of seashore vegetation in *The Old Margate Hoy* essay is applicable to some parts of an Arctic land: 'I hate those scrubbed shoots, thrusting out their starved foliage from between the horrid fissures of dusty innutritious rocks, which the amateur calls "verdure to the edge of the sea."' But in the scrubbed shoots of the Willows and the Dwarf Birch, with their profusion of catkins, doomed by the force of circumstances to lead a prostrate life on bare rock, on the faces of cliffs, or creeping among a miniature undergrowth of Moss, Lichen, and other plants, there is a beauty that arrests attention; and in the late summer, when the green leaves have turned to light orange or brilliant red and the Willow catkins are covered with open capsules releasing the white fluffy seeds, the ground becomes a mosaic of colour which would be difficult to match in many more favoured lands.

The influence of Lichens on colour production in nature is well illustrated in many parts of Greenland; one is reminded of Ruskin's description:

Far above, among the mountains, the silver lichen spots rest, star-like, on the stone: and the gathering orange stain, upon the edge of yonder peak, reflects the sunsets of a thousand years.... To them, slow-fingered, constant-hearted, is entrusted the weaving of the dark, eternal tapestries of the hills; to them, slow-pencilled, iris-dyed, the tender framing of their endless imagery.

At the small Settlement of Niakornat (Map *B*, N) the huts of the natives are built close to the beach or perched on ledges on the higher ground (Fig. 32). Seen from a distance the massive and partially rounded, though rugged, boulders and hills of volcanic breccia—a rock composed of angular pieces of a fine-grained and in part glassy lava embedded in a matrix of volcanic ash—produce a particularly gloomy impression by the contrast of their dark shoulders to the lighter hills near them; but on a nearer view the dark surfaces were seen to be almost covered with splashes of a vermilion Lichen. It is not improbable that, in the menacing headlands that guard the harbour of Niakornat and partially encircle the Settlement, we have the relics of a vast accumulated mass of ash and splintered rocks ejected from some old volcano in the immediate neighbourhood.

The peculiar construction of Lichens renders them less dependent than other plants upon the nature of the substratum on which they grow. As films of dull black they dapple the grey surfaces of the gneiss, while other species produce a harmony of orange, yellow, and grey. On stony ground, among bosses of protruding rock and mixed with prostrate or tufted shrubs of the heath vegetation, large cushions of grey Lichens which when dry crumble to the touch, the flat, deeply lobed fronds of a bright yellow species, and the clumps of erect branches of stouter forms, sometimes tipped with small scarlet balls, give light and brightness to the duller background.

The Botany of Greenland is too wide a subject for more than a passing reference in these pages; it is intensely interesting to the botanist not only because of the richness of the flora, but also on account of its past history,—the relation of the vegetation of to-day to that which preceded the Glacial period, and the routes by which the pioneers of the present plant population arrived.

A certain emotional influence is produced by the heath-covered hill-sides and swampy lowlands, by the scattered colonies of more brilliant flowers on the drier rock-strewn regions of this treeless land for the perception of which no knowledge of natural science is needed, and even the layman's sense of wonder is stirred when he considers what this display signifies as a triumph of the forces of life over adverse physical conditions.

It was impossible with the limited time at our disposal to obtain a representative collection of the seaweeds that occur among the rocks or on sandy beaches. The number of different kinds was comparatively small and the majority were familiar. Floating on the waters of the Davis Strait, the Vaigat, and elsewhere, and washed up on the shore, were innumerable Laminarias which had been torn from rocks below tide-level. The genus *Laminaria* includes large brown seaweeds that are common objects on our coasts, especially those with long flat ribbons or broader flattened expansions, like great leaves deeply cut into finger-like lobes, attached to the surface of submerged rocks by tufts of strong holdfasts at the base of long stems. The Greenland seas are characterised by the abundance and large size of these marine plants. A fairly large specimen of one of the commoner forms (*Laminaria longicruris*) is shown in the photograph taken at Godthaab (Fig. 33); the broad ribbon-like frond has strongly crinkled edges and, including the supporting flexible stem, it reached a length of several yards.

It has already been pointed out that a comparison of many of the plants obtained from the rocks in Greenland, more especially from Cretaceous strata, with plants which still exist in other parts of the world compels the inference that the Arctic vegetation of many million years ago resembled very closely, at least in some of its constituents, that of tropical or sub-tropical regions to-day. As we scan the impressions of leaves and twigs on the surface of a freshly broken slab of stone and allow the facts discovered to stimulate the imagination, the ice and snow disappear, the scrubby Willows and Dwarf Birch are changed into a forest of trees of many sorts, and luxuriant thickets of scrambling ferns replace the stunted heaths; Arctic Greenland is transformed into a tropical scene recalled from the past[1]. Similarly, pursuing the idea of contrasts and shifting our ground to a gravel pit in England, the discovery in a layer of peat, belonging to a geological period separated by thousands rather than millions of years from the present, of fragmentary though well preserved plants, many of which are identical with species still living on the ice-free fringe of Greenland, changes the familiar landscape into that of an Arctic land: these samples of a bygone flora[2] tell us that when England was under the influence of the Glacial period it was as Greenland is.

In Greenland the collector of fossils digs out of the rocks plants which he has gathered in the Tropics or species next of kin to them; in layers of peat among the deposits of lakes and rivers found among the Glacial gravels of England he sees again the flora of Greenland.

---

[1] An interesting summary of the evidence of climatic changes afforded by Arctic fossil plants was contributed by Dr Nathorst to the International Geological Congress of 1910: a translation of this by the late Dr Arber was published in the *Geological Magazine*, VIII, 217, 1911.

[2] For details see 'The Arctic Flora of the Cam Valley,' by Miss M. E. J. Chandler, *Quarterly Journal of the Geological Society*, LXXVII, 4, 1921.

PLATE XXIV

FIG. 40. SOUTH COAST OF UPERNIVIK ISLAND

FIG. 41. GLACIER, UPERNIVIK ISLAND

PLATE XXV

FIG. 42. UMANAK CLIFF

CHAPTER V

The key of the past, as of the present, is to be sought in the present; and, only when known causes of change have been shown to be insufficient, have we any right to have recourse to unknown causes.   HUXLEY.

More about fossil plants. The relation between rocks and scenery. Rocks formed as sediment under water and rocks that are volcanic. Dykes of basalt. The loneliness of a Greenland beach. Deserted Settlements.

MANY of the localities visited were on un-inhabited coasts where the land rose gradually inland for a few hundred yards; then the gradient rapidly increased up the face of the mountain. Deep ravines laid bare a succession of sedimentary strata 1000 ft. or more in thickness over which had been piled layer after layer of lava-flows and beds of volcanic ash (Fig. 34). Masses of frozen snow powdered with wind-blown sand or the darker dust from the shales were often met with in more shaded parts of the valleys. The widely spread sheets of basalt, in some places as many as thirty superposed layers, which give a terraced appearance to the weathered face of the cliffs like that on the rocks of Mull and other islands off the west coast of Scotland, are proof of long-continued volcanic activity on a stupendous scale. Another expression of the same kind of phenomenon is seen in the numerous dykes which frequently cut across the beds of sandstone and shale. A dyke consists of some igneous rock, often basalt, that has been formed by the slow cooling and crystallisation of molten material forced from below through cracks and fissures in the overlying strata. The softer rocks fall an easier prey to the action of the weather than the harder and more compact dykes which are left as great ribs or dark brown buttresses projecting on a light yellow background of less resistent material.

The photograph reproduced in Fig. 35, taken near Skansen, a small Settlement on the south-east coast of Disko Island, shows in the distance the basaltic plateau with an occasional glacier descending from the inland ice; in the foreground the sloping sides of the valley are formed of light yellow sandstones and beds of dark shale which at a higher level are covered and protected by volcanic rocks that were formerly continuous with the rest of the plateau. A cap of dark brown rock, part of a basaltic dyke, has saved the small hill of sandstone in the middle of the valley from the destructive action of weathering agents. Fig. 36 shows a much narrower zigzag valley eroded by ice streams through the sandy sediments; the cliffs in the foreground have been weathered into slender pinnacles and intersecting ridges; a little lower down the ravine a brown jagged dyke of basalt casts its shadow on the sloping surface of a smooth embankment of light yellow sand, and the sides of the farther hill are scarred with a succession of small gulleys. Beyond the lower heath-covered ground lies Disko Bay with scattered icebergs, and on the opposite shore, too far away to be seen in the photograph, are the hills of the mainland showing a strip of white against the sky where a depression in the coast-line reveals the edge of the inland ice. The dyke shown in profile on the edge of the sandstone slope in Fig. 36 is seen at close quarters in Fig. 37; it resembles a partially ruined, rusty brown wall on a wind-worn, denuded field of sand, the result of disintegration of beds of soft sandstone. Here and there, adhering to the sides of the dyke, are blocks of sandstone which were hardened and rendered more resistent to denuding agents by contact with the molten rock which welled up against their fissured sides.

For the most part the sandstones and shales are light yellow or dark grey in colour, but in some places the same beds exposed on the scarps of the hills are bright red, while others simulate yellow-white porcelain with clearly defined impressions or moulds of leaves and twigs. These red and yellowish white rocks, which ring under the hammer, owe their colour and texture to burning. They are sandstones and shales altered in all probability as the result of some spontaneous combustion.

Standing on the slopes of the hills one looks across from the south shore of the Nûgssuaq Peninsula to the mountains of Disko Island on the other side of the Vaigat—a distance of eight or ten miles though in the clear atmosphere it seems much less—the foot-hills with screes and talus-slopes surmounted by volcanic rocks, lines or patches of snow picking out the layered succession of lava-flow and ash. The identity in geological structure of the hill-tops on the two sides of the strait shows that the deep channel of the Vaigat has in comparatively recent times been cut through a vast plateau; initiated by some fracture in the earth's crust the channel was deepened and broadened by the action of denuding agents. The water is sprinkled with icebergs of all shapes and sizes, their bright whiteness contrasting with the dark blue slopes of the hills beyond which are relieved by the shining ice-covered summits and the bands of snow on the ledges of the lava and ash. Fogs in the Vaigat are fairly common. On one occasion looking down from the mountains of the peninsula the more distant part of the Vaigat was seen to be filled by a slowly advancing flood of white clouds, which gradually obliterated the icebergs and the bases of the cliffs until eventually the whole of the water and the lower ground was submerged and the rolling surface of the invading mist was illuminated above by a bright blue sky.

It was a comparatively rare event to see any signs of human life as we travelled along the Vaigat, but occasionally we met a solitary kayaker or an umyak. One evening a sturdy little Eskimo returning from hunting, the dead body of a seal made fast to the side of the kayak, paddled to our motor-boat. The square piece of white cloth at the bow of the kayak (Fig. 20) is a device employed when hunting to make the seal believe that the kayak is a piece of ice: a white cap worn by the hunter adds to the deception. We were assured that Polar bears when stalking seals sometimes push before them a block of ice. Being attracted by the native, we engaged him to accompany us for a few days and took his kayak on board. He soon became an expert fossil collector and expressed his satisfaction by singing doleful tunes that he had learnt in church. The natives generally regarded us with curiosity and, we were told, spoke of us as vagabonds or tramps.

Traces of former habitations in localities now deserted always appeal to the imagination, particularly on a desolate coast such as that of the stormy and gloomy Vaigat. A small wooden cross on the summit of a low hill close to the water's edge marked the burial-place of some nameless Greenlander from one of the half-ruined stone houses not far away and abandoned some years ago. On the site of the former Settlement there were luxuriant patches of two common grasses which are invariably found on ground rich in organic material in the neighbourhood of dwellings. Older graves were often met with, consisting of rough blocks of stone enclosing a small space in which, if the grave had not been disturbed, fragments of bones might still be seen. Another relic of the past, it may be of a comparatively remote past, was represented by the remains of two series of stones, placed on the ground about a yard apart and forming the intersecting arms of a cross twenty or thirty yards long. In former days the Eskimoes arranged blocks of stone in long rows to serve as a test of skill and endurance; the game consisted in the competitors hopping from stone to stone on one leg, the other being tied behind the back, and each competitor carried a seal on his shoulders. In an article on the Eskimo Stone Rows (Nangissats in the Eskimo language) contributed by Mr Porsild to the *Geographical Review* (vol. x, November, 1920) it is stated that they appear to be confined to localities between lat. 67° N. and lat. 72° N. and are unknown outside Greenland.

Walking alone on the shore one morning, far from any habitation, I saw with surprise, and with a suddenness that was startling, the figure of a dog silhouetted in the distance against the sky on the top of a large boulder. In the summer the dogs are usually left to themselves and it is not uncommon to meet them or to see their footprints on the sand far from a Settlement. To the Eskimo dog the contrast between winter and summer must be a very real one; in the winter the dogs are regularly fed and kept in good condition for the invaluable services they perform, but in the summer they become scavengers and vagabonds or are kept within a wired enclosure. They are only partially domesticated and are almost as much wolf as dog. Though generally not unfriendly, or at least indifferent, to human beings, they sometimes make furious attacks upon children or even adults. Hunger and wildness cause them to dominate the situation in the Settlements of North Greenland, where sledges are the only means of transport and travel in the winter; sledges and kayaks are kept on special stands raised several feet above the ground or on roofs of houses out of their reach.

CHAPTER VI

Some drill and bore
The solid earth, and from the strata there
Extract a register, by which we learn
That He who made it, and reveal'd its date
To Moses, was mistaken in its age.   COWPER.

Hare Island. Drift-wood. Upernivik Island. The catechist at home. Atanikerdluk; its scenery, fossil plants, and dykes. First and last impressions of Greenland.

ONE of the places it was our aim to visit was Hare Island, off the north-west coast of Disko Island, a small uninhabited island of basalt and beds of ash, including some layers with leaves and other fragments of a former vegetation which had been overwhelmed during volcanic eruptions. Landing on the beach is often very difficult on account of the swell, and it was only on our second visit that we were able to get ashore at the desired spot. On the north side of the island are the graves of some British sailors from whaling ships. Whalers are now very seldom seen: British ships no longer sail to these waters and in recent years they have been visited only by a few Norwegian vessels. In 1819 Sir James Ross on the way to the north saw as many as forty English whalers in the Vaigat. Hare Island is in some places rich in flowers, but wide expanses of dark brown basaltic sand with little or no vegetation give it a singularly desolate and depressing appearance. Some of the southern plants which occur on

Disko Island reach their northern limit on Hare Island, and these are regarded by Mr Porsild as relicts of a time subsequent to the Glacial period when the climate was more genial than at present. The Alpine *Potentilla* and the Alpine *Veronica* are two examples of the southern species recorded from Hare Island. These southern plants ripen their seeds and fruits late in the autumn and are met with only in places where they are well covered with snow during the winter.

On the beach were substantial pieces of drift-wood, a characteristic feature of many parts of the Greenland coast. In former days it was from logs of wood washed ashore that the Greenlanders obtained the whole of the timber used in the construction of their boats and weapons. A micro-scopical examination of samples of drift-wood has demonstrated that much of it comes from the forests of northern Siberia: after the fallen stems and branches reach the sea by Siberian rivers they are carried many hundred miles by the Polar current down the east coast of Greenland, round Cape Farewell and north again through Davis Strait and Baffin Bay. Drifted logs are abundant on the coasts of Spitsbergen, Bear Island, the small and lonely island of Jan Mayen, and other Arctic lands. If, as is often possible, the provenance of the wood can be ascertained, its distribution furnishes valuable information on the subject of ocean currents[1].

[1] An account of the botanical nature and provenance of the Greenland drift-wood was published in 1903 by Dr Ingrarson in the Transactions of the Royal Swedish Academy (*Kongl. Svensk. Vetenskapsakad. Hand.* vol. 37).

PLATE XXVI

E. Porsild phot.

FIG. 43. UMANAK CLIFF (LEFT) AND SAGDLIARUSEK, FROM KOOK

PLATE XXVII

R. E. Holttum phot.

FIG. 44. VIEW FROM BEHIND THE SETTLEMENT OF UMANAK

FIG. 45. ATANIKERDLUK
Basaltic dykes penetrating the sandstone; the largest forms
a V-shaped wall across the ravine in the centre

While most of the drift-wood is brought by the Polar current from the Siberian coast, some has been traced to Norwegian sources, and pieces of North American Pines and other trees have also been recognised. Currents from the south also transport plants to the Arctic regions as is shown by the occasional discovery of seeds of tropical plants on the West Greenland coast.

It was the discovery of pieces of wreckage from the American ship, the 'Jeanette,' which foundered north of the New Siberian Island in 1881 on the south-west coast of Greenland, and the study of the distribution of Siberian drift-wood that first led Nansen[1], and more recently other explorers, to trust to the motive power of currents as the chief factor in Arctic expeditions.

The logs of drift-wood stranded on a Greenland beach after a long sea voyage from their native habitat suggest the possibility that some at least of the stout petrified stems, which were found among the boulders in the beds of the glacial streams washed out of the sandstones on the hillsides, might in their day have been carried far from home, to be entombed with the waifs and strays of a contemporary Greenland flora to which they did not belong.

From Hare Island we visited Upernivik Island (lat. 71° N.), on the north side of Umanak Fjord. On the south coast there is a small Settlement at the foot of the hills commanding a view that it would be difficult to surpass for grandeur and

[1] *Farthest North*, 1897 (Fram Expedition, 1893–6).

harmony of colour. Fig. 38 shows a small portion of the view seen from the beach of Upernivik Island. The local catechist, a bespectacled Eskimo, who was my companion during a morning scramble over the rocks to a neighbouring glacier, appeared to be suffering from tuberculosis, a disease that is alarmingly prevalent in the country. After our evening meal on the motor-boat we decided to pay a visit to the catechist's house and took with us a supply of coffee berries from which an unusually tall and powerfully-built Eskimo woman prepared the very strong black coffee with a slightly burnt flavour which the natives love. There were only four cups: after the European guests had been supplied these were passed round in turn among two other groups of four. The room was small and without ventilation; callers dropped in one by one until the place was crowded with men and women, most of whom were smoking pipes, and small children. The walls were decorated with coloured pictures and post-cards, a mixture of pages from Danish comic papers and coloured prints of sacred subjects.

One of the houses of the Settlement was without a roof and appeared to be in ruins. Though to a less extent than formerly it is still a common practice in some districts for the Greenlanders in summer to lead a more wandering life, or to establish themselves in a good hunting locality, travelling in umyaks, and sleeping in tents made of skins of the bearded seal thrown over a cunningly arranged framework of poles and oars. On their departure from the more permanent Settlement the roof of the winter house is removed and nature is left to do the much needed spring-cleaning and airing. My intention was to take a photograph of a typical, roofless house and three Eskimo women were invited by signs to come into the picture. Unfortunately, so far at least as the prospect of obtaining a satisfactory photograph of the building was concerned, the whole female population formed up in line (Fig. 39).

From the beds of shale exposed in the cliffs in the foreground, shown in Fig. 40, above the beach littered with boulders, several impressions of fossil plants (Cretaceous) were collected, including many beautifully preserved leaves of *Ginkgo* (the Maiden-hair tree) and pieces of the large fronds of a Cycad, a plant related to the so-called Sago Palms, the majority of which flourish in the Tropics. An almost vertical dyke thrust through the old sediments forms a prominent feature on the left. The two massive mountains in the distance are portions of the highlands bordering the western edge of the Umanak Fjord, composed of some of the oldest rocks in Greenland.

On one part of the south coast of Upernivik Island the heterogeneous collection of boulders, which are the products of ice-action when the climatic conditions were more severe and the inland ice was greater in extent and glaciers were larger, presented interesting geological problems.

A short distance from the Settlement on Upernivik Island the strata of sandstone and shale are separated from the infinitely more ancient rocks of which the greater part of the island is composed by a valley occupied by a fairly large glacier (Fig. 41), on the flanks of which there are accumulated masses of lateral moraines. The present snout of the glacier is separated from the sea by a few hundred yards of sloping ground formed by the terminal moraine; but within the memory of living man, so we were told, the ice reached the sea. Near the glacier is a sandy beach where clumps of Willow, the blue-green leaves and flowering stalks of *Elymus*, a common dune-forming grass on the coasts of Europe, and spreading colonies of the succulent foliage-shoots of the Sea Purslane (*Arenaria peploides*) formed the nuclei of miniature sand dunes. An occasional Snow Bunting, inquisitive Gulls, a few brown Butterflies, and swarms of aggressive Mosquitoes kept me company as I lay on the sand looking across the fjord at the hills of the Nûgssuaq Peninsula, stretching to the west, and in front of them, towards the eastern end, the conical Umanak cliff (Fig. 14; the hill in line with the bowsprit of the motor-boat), in shape recalling the Matterhorn. When we afterwards gained a nearer view of this rock, its dignified isolation made a deep impression on my mind: it rises almost sheer from the sea, a wall of crystalline igneous rock nearly 4000 ft. high (Fig. 42). The salmon-pink mass is cut across half way up the precipice by a thin black band bent on itself like an S lying on its side (Fig. 43), an eloquent witness to the intensity of the forces which folded and

crumpled the rocks of which the Umanak mountain remains as a detached and magnificent bastion
that in the course of ages has been fashioned into
its present form. The photograph reproduced in
Fig. 43 shows the view looking north-east from
Kook, an uninhabited part of the north coast of
the Nûgssuaq Peninsula famous for fossil plants
(Cretaceous); it lies to the south-west of Umanak
where the dotted line bends towards the coast
(Um., Map B). The mast of the motor-boat 'Angut'
is between the cliff of Umanak, on the face of
which the reversed fold of black schist is faintly
seen sloping towards the right, and the bold
headland of the island of Sagdliarusek composed
of a pink crystalline rock. The Umanak cliff
is five miles away and the island to the right
nine miles. Near the base of the cliff is the Settlement of Umanak (Fig. 44) with the native huts
and timber houses built on the small level patches
among the rounded hummocky rocks which form
a massive plinth to the pyramid that dominates
the island. A motor-boat is at anchor in the
sheltered bay and beyond the smooth water of
the fjord, sprinkled with glistening bergs, is the
massive side of Sagdliarusek Island, a part of which
is seen also in Fig. 43. As we lay at night off the
coast the howling of dogs in chorus as they prowled
over the rocks and the recurrent boom of breaking
icebergs seemed in harmony with the spirit of the
place. On the following day we were suddenly
transported to civilisation. The Inspector for
North Greenland arrived in his motor-boat and

having put on the official uniform, including a cocked hat, he pinned the badge of knighthood on the breast of the Danish Factor and presented medals to a few of the senior natives. This ceremony would have been performed by the King of Denmark had not the disaster to the S.S. 'Bele' interfered with his plans. Immediately after the presentation the Inspector took snapshots of his audience; the King's representative became an interested photographer. Coffee and cigars were handed round to the men and women of the Settlement and at the conclusion of the more public ceremony the Danish officials and their English guests toasted the recipients of the royal honours in official wine.

The scarcity of good harbours on the coast of the Nûgssuaq Peninsula and the sudden storms are serious drawbacks to travellers who trust solely to a motor-boat and have not a supply of petrol sufficient to enable them to return to a place which they have been compelled, unexpectedly, to leave through stress of weather. The proximity of icebergs on an exposed coast is also a source of danger; frequently during the night blocks of ice and small icebergs bumped against the sides of the boat and had to be dealt with promptly.

One of the best known localities in Greenland for fossil plants is Atanikerdluk (Map *B*, Atk.), an uninhabited part of the coast separated by a wide stretch of sandy beach from a peninsula that juts out into the Vaigat. It was here that we had our greatest disappointment in the course of a four-weeks expedition. On the evening of the second day the two natives on board assured us that bad weather was imminent and very reluctantly we left with our work half done. Shortage of petrol rendered our return impossible. In the face of a cliff rather more than 100 ft. above sea-level, there is a dark band of shale (an old mud rich in the products of decay of Cretaceous plants), which contains impressions of the large fronds of a Cycad, many twigs of Conifers allied to species that are now confined to much warmer climates, and a considerable variety of broad-leaved trees. One of the most interesting of the broad-leaved forms was identified several years ago as a species of *Liriodendron*, a genus now represented by the Tulip-tree, a native of North America and China, which is often seen in cultivation in English gardens. In addition to the bed containing leaves of a Tulip-tree there are several other exposures of rocks at Atanikerdluk from which many specimens of Cretaceous and Tertiary plants have been obtained, but our unexpectedly sudden departure made it impossible for us to secure a representative collection.

At some localities the number and quality of the specimens of fossil plants collected exceeded expectation, while at others the results were disappointing. At Atanikerdluk, owing to circumstances beyond our control, one felt that the best material was being left in the rocks:

> The worst we stored with utter toil[1],
> The best we left behind!

[1] It is only fair to add that my companion bore most of the toil of packing, and with good results.

The main ravine at Atanikerdluk (Fig. 45) is not only of special interest geologically; it also affords a most remarkable display of dykes and illustrates on a grand scale the relation between scenery and rock structure. A stream flows among jumbled heaps of boulders at the bottom of a steeply inclined valley; the valley slopes consist of natural embankments of loose, light yellow sand mixed with milk-white rounded pebbles of quartz —detritus formed by the erosion of the sandstones which are here the dominant rocks—and in places the talus slopes are replaced by exposures of the rocks themselves, thick beds of sandstone with no division into layers, thinner, well-stratified beds, bands of shale, and an occasional seam of coal. These sedimentary strata, having a total thickness of many hundred feet, exhibit here and there on an exposed plane of bedding a series of ripple-marks, and afford other evidence of their origin as sheets of sand and mud in shallow water and among drifting currents. Many of the sandstones are made up of thin layers, often rendered more conspicuous by the presence of iron-stained bands, which exhibit the well-known arrangement spoken of as current-bedding: a series of layers sloping at a certain angle is cut off by another set sloping in a different direction. This frequent variation in the lie of the thin beds is evidence of the deposition of the sandy sediment in water with eddying and shifting currents. Dark brown dykes cut across the sands and sometimes intersecting dykes project like part of a huge network; but the most im-

PLATE XXVIII

FIG. 46. DYKE IN THE MAIN RAVINE, ATANIKERDLUK

PLATE XXIX

FIG. 47. A TABULAR ICEBERG

pressive example of intrusions of igneous rock is afforded by vertical walls of basalt which stretch across the valley. These magnificent dykes do not form continuous curtain walls from one side of the valley to the other, but the rocks are twice or thrice stepped on each side; the light brown wall of basalt towers against the sky at least 100 ft. above the upper level of the slope of the ravine. Its jagged and weathered ledge projects horizontally for some distance towards the middle of the valley and is then cut vertically down into a deep step, and this is repeated two or three times. At the foot of the valley the dyke crosses the stream as a resistant barrier where the water falls in a cascade.

The ravine at Atanikerdluk (Fig. 46) stirred the imagination more than most of the many impressive scenes in other parts of Greenland. To anyone interested in geology a sense of sharp contrasts between the present and the past is so constantly evoked by the interpretation of the rocks that the wonder of it becomes less intense; but there are places and circumstances in which this sense of change from one age to another is awakened with especial force and vividness. The icebergs floating on the arm of the sea, a thousand feet below the ravine, are fragments of the shield of ice that has lain for thousands of years over nearly the whole of the interior of the country; the clumps of Arctic plants on the sloping banks of weathered sandstone on either side of the glacial stream are in harmony with the climatic conditions of the present moment of geological history. The waifs and strays of a lost vegetation contained in the sandstones and associated shales suggest a scene very different from the present—a time when the rocks of the valley were slowly accumulating, as gravel, sand, and mud, in the delta of a river flowing between banks clothed with a rich and varied vegetation. In Greenland to-day the vegetation lives precariously on the rocks which were uplifted from the water some millions of years ago. Higher up the ravine lavas and beds of ash spread in a succession of layers over the sedimentary rocks recall a period of intense volcanic activity and, most eloquent of all, the towering weather-beaten walls of basaltic dykes compel the mind of an observer, familiar with ordinary geological evidence, to picture fissures riven in the strained sandstone filled with molten basalt welling up from a subterranean reservoir. These glimpses of the past and their disharmony with the present impart to the reality of geological history a sense of unreality, an impression that may best be expressed by one of the conceptions of Milton:

> Thither, by harpy-footed Furies haled,
> At certain revolutions all the damn'd
> Are brought; and feel by turns the bitter change
> Of fierce extremes, extremes by change more fierce,
> From beds of raging fire to starve in ice
> Their soft ethereal warmth, and there to pine
> Immovable, infix'd, and frozen round,
> Periods of time; thence hurried back to fire.

On the left-hand side of the cliff rising from the beach (Fig. 45) a dyke cuts obliquely across the almost horizontal Cretaceous sediments, while other

ridges of intruded basaltic rock ascend the higher
slopes, one of them dipping steeply down into
the ravine in front of a light yellow face of rock
which reflects the light: a continuation of the
same dyke rises as a lofty curtain wall on the right-
hand slope of the valley. Fig. 46 represents an
attempt to convey something of the impression
made by the weathered and towering masses of
the rust-coloured basalt dyke penetrating the
eroded slopes of the sandstone above the stream.

The light yellow tones of the sandstones and
the darker tones of the basalts are relieved by
clumps of bright yellow Dandelions and Arnicas,
purple Willowherbs, and clusters of the tall russet-
brown *Oxyria*, a plant allied to the common Dock.

From Atanikerdluk we went to the small Settlement of Sarkak on the mainland, where the
gneissic rocks replace the sedimentary strata and
basalts of the Nûgssuaq Peninsula. From Sarkak
a short excursion in search of fossils was made in
an umyak to a place affording a good view of the
Sarkak valley in which flows the largest river we
saw. On one side of the broad valley are hills of
gneiss with the much more modern flat-topped
basaltic hills beyond; on the opposite side the hills
consist of the sandstones and other sedimentary
rocks we had seen at Atanikerdluk. In the valley
itself dark masses of intruded igneous rocks form
conspicuous bosses and dykes which lie above the
river plain as huge serpentine ridges. At a height
of 1400 ft. on the sandstone hill-side the familiar
and hardy Harebell (*Campanula rotundifolia*), which

has wandered over the mountains of Central Asia, Japan, and North America, Saxifrages, and several other flowering plants were in full bloom.

On the way from Sarkak to the Settlement of Ritenbenk, on an island off the mainland, we passed near the vertical face of a high cliff known as the Bird Rock. It was nearly midnight in the middle of August. The smooth sea was a rich indigo, and between our boat and the light sky over the horizon was silhouetted a small trading schooner which we had taken in tow. As we approached the Bird Rock, pairs of Razorbills floating on the water became increasingly numerous, always a large and a smaller bird, mothers giving lessons in diving to their inexperienced children. An Eskimo on board the schooner unable, as are the natives generally, to resist the temptation of firing at any wild thing, tried to shoot the swimming birds; the mothers sometimes dived to avoid the shot, leaving the young birds to their fate, but usually the maternal instinct prevailed and the older birds with cries of alarm kept by the side of their charges.

From our first sight of Greenland until we left for Copenhagen the days were always interesting, whether we were slowly travelling from place to place collecting specimens or living at the Arctic Station botanising and geologising in the immediate neighbourhood, and learning many things about the people and the country from our good friend Morten Porsild, the Director. My first introduction to this little-known land and the last view of the receding mountains of the south coast on our return voyage remain as permanent records of the beginning and end of a memorable summer in Greenland.

We had our first sight of Greenland late in the evening, about an hour before midnight, at close quarters: the sun had set, we were still south of the Arctic Circle: on each side of the ship were rocky islands and promontories with occasional patches of snow on their flanks, and the summits hidden by clouds. Beyond the headlands in gradually deepening gloom were faintly outlined hills receding into the arms of a fjord. On passing a part of the coast where we were told was a small fishing settlement the steamer, with a suddenness that seemed to break the spell, gave two shrill whistles and soon afterwards a solitary Eskimo was dimly seen as a black speck vigorously paddling his kayak in the hope of overhauling us and probably expecting to be taken on board. This was a fitting introduction to a land where the very existence of the inhabitants depends upon their skill in the use of the incomparable kayak.

It is comparatively seldom that a passenger in one of the Danish government steamers has the good fortune to see Cape Farewell or the adjacent parts of the coast. The abundance of ice and the prevalence of fogs and storms necessitate a course well out to sea. On our outward voyage Cape Farewell was passed at a distance of about thirty miles in the middle of the night: a violent storm of hail and wind and the thick weather compelled the captain to lay to for some hours.

On the return journey the weather was more favourable and we had a wonderful view of the coast line near the southern extremity of Greenland. On the horizon forty or fifty miles to the north we saw a jagged line of Alpine peaks, some tapering to slender conical points, others having the form of more massive pyramids separated from one another by depressions which seemed to show against a greenish blue band of sky glimpses of the inland ice; though it may be we were looking along arms of some of the tortuous fjords that cut deep into the coast. The light of the sea contrasted with the deep blue of the mountainous headlands against a pale steely-blue background cut off by an over-hanging bank of dark cloud. Later in the evening the clouds dispersed and the serrated profile of the mountains was sharply outlined against a luminous sky; the 'golden splendour of the north' faded into night. The rapidly changing scene produced an impression of sadness and majesty; it was our farewell to a land which in some aspects merits the name given to it more than three hundred years ago—the Land of Desolation; it is a land remarkable for the splendid dignity of its scenery and possessed of a subtle power of inspiring affection tempered by a sense of awe.

INDEX

CAMBRIDGE : PRINTED BY J. B. PEACE, M.A., AT THE UNIVERSITY PRESS